What is a p-value anyway?

34 Stories To Help You Actually Understand Statistics

Andrew Vickers
Memorial Sloan-Kettering Cancer Center
Department of Epidemiology and Biostatistics

Addison-Wesley

Boston   Columbus   Indianapolis   New York   San Francisco   Upper Saddle River
Amsterdam   Cape Town   Dubai   London   Madrid   Milan   Munich   Paris   Montréal   Toronto
Delhi   Mexico City   São Paulo   Sydney   Hong Kong   Seoul   Singapore   Taipei   Tokyo

Editor in Chief: Deirdre Lynch
Acquisitions Editor: Christopher Cummings
Assistant Editor: Christina Lepre
Editorial Assistant: Dana Jones
Senior Managing Editor: Karen Wernholm
Senior Production Supervisor: Peggy McMahon
Senior Design Supervisor, Cover and Interior Design: Andrea Nix
Chapter Opener Illustrations: Matt Andrews
Marketing Manager: Alex Gay
Marketing Assistant: Kathleen DeChavez
Senior Author Support/Technology Specialist: Joe Vetere
Manufacturing Manager: Evelyn Beaton
Senior Manufacturing Buyer: Carol Melville
Production Coordination, Technical Illustrations, and Composition: Laserwords Maine

Photo credit: p. 148, Portrait of the English statistician and geneticist, Sir Ronald Aylmer Fisher (1890–1962), © Photo Researchers, Inc.

Many of the designations used by manufacturers and sellers to distinguish their products are claimed as trademarks. Where those designations appear in this book, and Pearson was aware of a trademark claim, the designations have been printed in initial caps or all caps.

Library of Congress Cataloging-in-Publication Data

Vickers, Andrew
  What is a P-value anyway? : 34 stories to help you actually
understand statistics / Andrew Vickers.—1st ed.
      p. cm.
 ISBN 0-321-62930-2
 1. Mathematical statistics. I. Title.
QA276.12.V53 2010
519.5—dc22

                          2009016210

Addison-Wesley
is an imprint of

www.pearsonhighered.com

ISBN-13: 978-0-321-62930-2
ISBN-10: 0-321-62930-2

Dedication

This book is dedicated to four mentors who taught me not only how to do statistics but, more importantly, why.

Iain Chalmers

David Sackett

Doug Altman

Colin Begg

About the Author

Andrew Vickers is an Associate Attending Research Methodologist in the Department of Epidemiology and Biostatistics at Memorial Sloan-Kettering Cancer Center in New York. He is active in a variety of fields of cancer research, including surgical outcomes, molecular markers, and clinical trials. He also conducts original research in statistical methods, particularly with respect to the evaluation of prediction models. At the time of writing, he has been principal author or coauthor on over 200 peer-reviewed scientific papers.

Dr. Vickers has a strong interest in teaching statistics. He is the course leader for the Memorial Sloan-Kettering Cancer Center biostatistics course and teaches biostatistics to medical students at Cornell Medical School. Dr. Vickers lives with his wife and children in Brooklyn, New York.

Contents

Hypothesis Testing

Regression and Decision Making

Some Common Statistical Errors, and What They Teach Us

How to read this book

It is an odd feeling when you love what you do and everyone else seems to hate it. I get to peer into lists of numbers and tease out knowledge that can help people live longer, healthier lives. But if I tell friends I get a kick out of statistics, they inch away as if I have a communicable disease.

I have started to think that most folks' views of statistics as a refined type of torture go back to how it is taught, and to textbooks in particular. Statistics textbooks can be long, boring and expensive. With this in mind, I proposed to my editor that I write a book that was short, boring and expensive. He considered it but eventually decided I needed to come up with something better. So I thought about it this way: the typical statistics textbook (a) tells you how to do statistics, not how to understand it, (b) is full of formulas and (c) is no fun at all. I wondered whether I could write something that (a) focused on how to understand statistics, (b) avoided formulas and (c) was fun, at least in places.

This is how I came up with the idea of stories. The 10th commandment is "You shall not covet your neighbor's house, wife, donkey, or ox" but no one says this in conversation. Instead, they say "the grass is greener." In case you didn't know, "the grass is greener" comes from an old story about some goats that were happily eating grass in a field until they looked up and noticed the grass on the other side of a small stream. The grass looked so much greener there that they crossed over a little bridge. But after feeding for a while, they looked up again and thought that, actually, the grass on the other side of the stream, back where they had started, looked a lot greener than in the field where they were standing. And so they spent the day crossing the bridge back and forth, always thinking that the grass was looking greener on the other side. I think the last time I heard this story was in kindergarten, but I still remember it and what it means. The 10th commandment is spot on but is hard to remember because it tells you what to do; you hear a story to help you understand something and you'll remember it for life.

Like stories, the chapters in the book are intended to be short and fun to read. The second half of the book, the discussion section, is a little weightier. The discussion questions vary: there is usually one question, the first, that is pretty essential and is something that you should really try to think about. Most of the others could be considered optional—some are there only for the really enthusiastic types (I flagged these). For example, there is a discussion on the derivation of a mathematical constant called e and an introduction to statistical programming.

If you have some experience with statistics, feel free to dip in and out of the book. Otherwise, you should probably try to read the chapters through from beginning to end. The first 12 chapters deal with some basics, such as averages, variation, distributions and confidence intervals. I then have a few chapters on hypothesis testing and p-values, before discussing regression—the statistical method I use most in my work—and decision making—which generally should be, but often isn't, what statistics is about. The last third of the book, starting from the chapter "One better than Tommy John", is devoted to discussing a wide variety of statistical errors. If it seems odd to devote so much of a book to slip-ups, it is because I have a little theory that "science" is just a special name for "learning from our mistakes." When I teach, I give bonus points for any student giving a particularly dumb answer because those are the ones we really learn from. In fact, I don't think you can really understand, say, a p-value, without seeing some of the ways it has been misused and thinking through why these constitute mistakes. So please don't blow these chapters off thinking you've read the stuff you'll be examined on: the final chapters will really fill in your statistical knowledge.

What this book can and can't teach you

Hopefully, after you have read this textbook, you'll have a good understanding of many of the key ideas of good statistics. I also hope that you'll be able to avoid some of the most common statistical mistakes and errors.

What you won't know how to do is actually do any statistical analyses, in short, because I haven't provided any of the appropriate formulas. If you want to conduct analyses for your research or for your coursework, you'll have to look it up a conventional statistical textbook with formulas and step-by-step instructions. Also, the book won't be particularly useful as a reference textbook to look up things that you've forgotten. So if you want to run statistical analyses, this should not be the only book you buy. (Although it should be the only book you buy multiple copies of, to give to your friends, family, colleagues, neighbors and random people you meet.) On the other hand, if you are the sort of person who doesn't want to do any statistics yourself—which is, I guess, most of the world—but have to understand and interpret statistics that you read—which is more of us than you might think—then this book might well be all that you require.

Where is the section on design?

I am a very design oriented statistician. As a quick example, missing data is a big problem in medical research. Statisticians have written hundreds of research papers proposing complex statistical techniques that predict what the data would have been, had it not been missing. My own contribution was to propose a very simple technique to reduce the rate of missing data in the first place, which is to telephone patients at home and ask them just two questions in place of a long questionnaire. In this way, we reduced the rate of missing data in a trial from 25% to 6%, which made the use of complex missing data analyses rather redundant.

As such, you might be surprised that there is no section on design in this book. In short, this is because I don't think you can separate out design from the rest of statistics and have a special chapter on it. I have two different chapters on regression analysis and the Wilcoxon test because, in theory, you could do one without the other; you can't think about either the Wilcoxon or regression analysis without considering the design of the study you are analyzing. Accordingly, I don't have a chapter on design. Instead, comments on design are woven throughout the text.

About the stories and data in this book

When I started writing, my editor said to me, "Andrew, I want you to write the funniest statistics textbook ever!" So I thought, "Great, I'll write one joke and then I'll be done."

Actually, it didn't quite happen exactly like that, but it isn't far off. On which point, the stories and data in this book were developed to help you learn statistics. This has sometimes meant simplifying or altering something to make it easier to understand. In some cases I simulated data ("simulation" is statistics speak for making stuff up). I did so on the grounds that the data I had to hand were much too complicated and would take far too much explaining and, as such, would detract from the reason I wanted to use the data in the first place—which was to help you to understand something about statistics. Also, you'd get sick of hearing about prostate cancer, which is the main thing I study.

Accordingly, the stories and data that follow are not all 100% factually accurate. I don't think I have said anything misleading, but please don't use the book to come to conclusions about blood counts in Swedish men (see *Chutes-and-ladders and serum hemoglobin levels: thoughts on the normal distribution*), prostate cancer (see *When to visit Chicago: About linear and logistic regression*), how long it takes for African-Americans to hail a cab (see *Some things that have never happened to me: Why you shouldn't compare p-values*) or, for that matter, my friend Mike (see *Regression to the Mike: A statistical explanation of why an eligible friend of mine is still single*). Or even whether "scared straight" helps juvenile delinquents avoid a life of crime (see *The probability of a dry toothbrush: What is a p-value anyway?*): it doesn't, and I say it doesn't, but don't take my word for it, look it up for yourself (see www.cochrane.org). This is, after all, a book about statistics, not crime policy.

I did analyze data sets for this book and present, without fudging, the results I found. You should be able to replicate my analyses. Much of the raw data is available on the web, but if you can't find it and want to replicate something, please let me know and I'll see how I can help. Incidentally, for most categorical data analyses in this book, I used Fisher's exact test.

I would like to acknowledge the Pew Research Center (www.pewresearch.org), which publishes raw data from its fascinating surveys of the American public. The data on attitudes to marriage between religions were adapted from the Northern Ireland Life & Times Survey 2006 (www.ark.ac.uk). The US 1996 crime statistics are available from www.statcrunch.com, an excellent resource for data sets for teaching (although, unlike the other data sets mentioned here, this is available only by subscription). The acupuncture and headache data set can be downloaded from www.trialsjournal.com/content/7/1/15 (where you can also read some of my thoughts about data sharing). The data on prostate cancer (and blood counts in Swedish men) come from a series of studies I have been conducting with my colleague, Hans Lilja. You can find out more by searching the medical database "PubMed" (http://www.ncbi.nlm.nih.gov/sites/entrez) for "Vickers Lilja". The data on maternity leave come from the work of Janet Gornick (see, for example, *Families That Work: Policies for Reconciling Parenthood and Employment*. New York: Russell Sage Foundation, 2003).

Acknowledgments and thanks

I'd like to thank Carol Peckham and David McNeel at Medscape, and Chris Cummings at Pearson, for being prepared to think a little differently; Dana Jones at Pearson for making it happen; Barrie Cassileth, for giving me the opportunity to work at MSKCC; Peter Scardino, for creating the environment where I could flourish; Mike Kattan, for leaving town; Mike V. (and numerous others), for being a good sport; Mike Shapiro for help with the cartoons; Burkhard Bilger, for advice on writing; Elaine McDonald-Newman, Jackie Miller, Bill Rayens, and Jeanne Osborne, for supportive feedback as well as constructive criticism; my colleagues at MSKCC in biostatistics and health outcomes, who have taught me more than I can remember; my mother-in-law, on the grounds that there are otherwise no mother-in-law jokes in the book; Angel Cronin, Caroline Savage, and Ally Maschino, for their passion, smarts and incredible productivity; and Caroline, Oliver, Emmeline, and Robin, for reasons that are obvious.

I'd also like to thank the following reviewers, who commented on this text in its early stages: Michael Dohm, *Chaminade University*; Sharon Testone, *Onondaga Community College*; Malissa Trent, *Northeast State Technical Community College*; Elizabeth Walters, *Loyola College in Maryland*; Sheila O'Leary Weaver, *University of Vermont*.

One of the jokes in Chapter 1 was conceived, written and directed by Emily Tobey, with additional help from Emily Tobey. The executive producer was Emily Tobey.

Medscape

Some of the material in this textbook was initially published as part of the *Medscape* series *Stats for the Health Professional*. *Medscape* offers the web's largest collection of free, full-text, peer-reviewed clinical medicine articles, along with news, views, free CME and multimedia expert opinion. *Medscape* also has features on specialist sites by topic areas—such as urology or gastroenterology—and professional discipline—such as nursing or pharmacy. Access to *Medscape* is free of charge, after registration, at www.medscape.com.

I tell a friend that my job is more fun than you'd think: What is statistics?

I have a friend who is a well-known DJ. He was once telling me about a gig he'd recently played in Milan—some kind of massive party for 20,000— when a neighbor of ours rolled up in a cab. A journalist, he was just returning from the Pacific Northwest where he'd been researching a story on wild mushrooms. The three of us stood chatting on the sidewalk for a while, and I thought I should mention that I was going to Cleveland to give a talk about bladder cancer.

I think they were impressed.

The DJ's unnaturally supportive wife once asked what it was that I did all day. What I said was: it's a lot of hard work and although it isn't as much of a thrill as DJ'ing a rave, it is more fun than you might think and is very satisfying. Best of all you get to meet a lot of other statisticians (actually, I didn't say that). This sounds like a bit of a non-answer, but as I'll explain, what I said made reference to *inference* and *estimation*, which is pretty much the A–Z of statistical analysis.

Estimation is about trying to work out how large or small something is. That "something" generally either can't be directly measured, or would take too much time and effort to do so. The two estimates in my answer were: "statistics is a lot of hard work" and "being a statistician is very satisfying." Ok, not very precise, but we might imagine that some psychologist had developed a questionnaire measuring mental effort, job satisfaction and, while we are at it, fun. To answer our question, "How much work is being a statistician?", what we'd ideally do is send the questionnaire to every single statistician in the world. But that would be kind of a pain, so we'd probably be better off sending the questionnaire to, say, 500 statisticians and hope that their answers were representative of statisticians in general. Let's say that our sample of 500 statisticians scored an average of 88% on the "mental effort at work" questionnaire: 88% is then our estimate of the average for all statisticians.

Inference is about drawing conclusions, and statisticians usually make inferences by testing hypotheses. My answer to the DJ's wife included two hypotheses: "doing statistics is not as much of a thrill as DJ'ing" and "statistics is more fun than you might think." To test the first hypothesis, we could give our "fun at work" questionnaire to 500 DJs and then compare their answers to those of the 500 statisticians (ok, it probably isn't worth doing this); to test the second hypothesis we just compare the statisticians' answers to some guess we'd made before we'd taken the data (e.g., statistics sounds like it would be about 0.3% fun).

This just leaves the issue of how we choose the statisticians to sample, which questionnaire we give them, whether we put a stamp on the return envelope or use "postage paid," how often we should chase up the ones that don't get back to us, how we enter data from all 500 questionnaires onto a computer, what you do if someone only fills in half of the questions, and so on and so forth. *Designing* a study is pretty complicated and statisticians know a lot about study design: it is said that R A Fisher, one of the great statisticians, used to clean out the rats' cages himself on the grounds that "if the rats are dirty, so will my data be." Indeed, all of the questions about the design of the questionnaire study (even whether you should put a stamp on the return envelope) have been written about by statisticians.

Given the choice, I guess most of us (me included) would rather be eating wild mushrooms, or spinning discs at a party, than running a complex statistical analysis. That is, until someone we love gets bladder cancer and we want to know what to do about it. Statistics sounds pretty cold—what is a number to me is somebody's living and breathing father, with stories to tell—but it has a very human goal: we want to live our lives better. To do that, we have to make good decisions and sometimes looking at numerical data in the right way can help us do so.

• **Things to Remember** •

1. Statistics involves estimation, inference and study design.
2. Estimation is about trying to work out how large or small something is.
3. Inference is about drawing conclusions, usually by conducting a statistical test of a hypothesis.
4. A hypothesis is a statement about the world that could be tested to see whether it is true or false.
5. Many studies produce numbers; as experts in numbers, statisticians often have a lot to say about how exactly a study should be designed.
6. Cleveland gets a bad rap, but the faculty dinner I had was actually pretty good.

for Discussion

1. I defined a hypothesis as "a statement about the world that could be tested to see whether it is true or false." Are there some statements that can't be tested?

2. There are two sorts of estimates that statisticians make: how big or small something is and how big or small something is compared to something else. An example of the first sort of estimate is "the mean height of an American male is close to 5 ft 9 $\frac{1}{2}$ in." An example of the second sort of estimate is "men who smoke are 23 times more likely to develop lung cancer than men who have never smoked." Write down some examples of estimates of both sorts.

3. Most hypotheses can be rephrased in terms of estimates. I mixed up some estimates and hypotheses below. Match each estimate with the corresponding hypothesis and say which is the estimate and which the hypothesis.

   a. Crimes decreased 21% comparing the year before and the year after completion of a program to improve street lighting.

   b. Men and women do not differ in their voting behavior for presidential candidates.

   c. Obesity rates in California increased during the 1990's.

   d. Recurrence rates were 5% lower in women receiving chemotherapy after surgery compared to women receiving surgery alone.

   e. The proportion of women voting for Democratic presidential candidates is 5% higher than men.

   f. Improvements in street lighting decrease crime.

   g. Electric shocks (punishment) are more effective than sugar (reward) for improving learning in rats, as measured by time to complete a maze learning task.

   h. Mean time to complete a maze was 20 seconds shorter in rats exposed to shocks than those given sugar.

   i. Obesity rates in California almost doubled between 1990 and 2000, from 10% to nearly 20%.

   j. Chemotherapy plus surgery is no more effective than surgery for breast cancer.

4. Who said "there are lies, damned lies and statistics"?

NOTE: See page 154 for answer sets.

CHAPTER 2

So Bill Gates walks into a diner: On means and medians

A statistician's joke: So Bill Gates walks into a diner … and the average salary changes.

Ok, not very funny, I realize.

The point of the Bill Gates "joke" is to illustrate the difference between two different types of average, the *mean* and the *median*. Let's imagine that the salaries in the diner before Bill walked in were as follows:

Eric	$85,000
Jose	$50,000
Barrett	$45,000
Sandra	$40,000
Todd	$35,000
Michael	$30,000
Katie	$30,000

The mean is what we normally think of as the average: you add up all the salaries and divide the total by the number of people. If you add up all the salaries ($315,000) and divide by the number of people (7), you get $45,000. The median is best thought of as the "middle" number: line up all the salaries from lowest to highest; the median salary is the one halfway along. The middle of 7 is 4, so the median salary in our diner is that of Sandra, who has the fourth highest salary, $40,000.

So now Bill Gates walks in with an annual income of, say, $1 billion (most people would call this rich, statisticians call it an *outlier*). Bill's salary changes the mean salary to a little over $125m. As for the median, there is no middle of 8, so we go halfway between 4 and 5. The 4th highest salary is now $45,000 and the 5th is $40,000, giving a new median salary of $42,500. Most people would say that $42,500 was a fair reflection of the salaries in the diner and that $125m had nothing to do with anything. And so we end up with a neat little rule: if there are outliers in the data—which is exactly what happens whenever a major software entrepreneur feels like having a greasy breakfast—use a median.

Here is the key point: if you stumble across a "neat little rule" in statistics, be very careful. Wanting a "fair reflection" of the data is not the only thing we want to use a statistic for; the other is to plan and make decisions. So let's imagine that instead of a diner we had a hospital, and instead of salaries, we had costs of surgery; in place of Bill Gates, we have a patient who has a series of complications after surgery, leading to costs of $250,000.

Patient	Cost of surgery
1	$85,000
2	$50,000
3	$45,000
4	$40,000
5	$35,000
6	$30,000
7	$30,000
8	$250,000

This gives a mean of just over $70,000; the median is the same as the Bill Gates example, $42,500. Which number would be most important if you were, say, a hospital administrator? $42,500 may well be a "fair reflection" of the typical cost of a patient's care, but writing next year's budget assuming costs of $42,500 per patient will likely lead to a shortfall. Thinking about means and medians is also why I buy health insurance (the median yearly expenditure of Americans like me is far less than the premium but when I look at mean yearly expenditure, I reckon I get a pretty good deal) and wear a seat-belt (even though the median number of injuries per car trip is zero).

So next time you are in a diner, and are bored and depressed because Bill Gates hasn't shown up yet, here is a neat little rule to scrawl on your coffee-sodden napkin: sometimes there are no right and wrong statistics; it all depends on what you want to use them for.

> ### • Things to Remember •
>
> 1. What most people call an "average" is what statisticians call a mean. To calculate a mean, think of your data as a list of numbers, add up all the numbers and then divide by the number of items on your list.
> 2. The median is the half way point of your list of numbers: half of the sample have values higher than the median and half have values lower than the median.
> 3. An outlier is when you have an observation that doesn't follow the pattern of the data.
> 4. When you have outliers, the median often gives a fairer reflection of the data than the mean.
> 5. Generally speaking, means are better than medians for planning and making decisions.

for Discussion

1. I said that "half of the sample have values higher than the median and half have values lower than the median." Is that always true?

2. Here is a die rolling game: you roll a die and if you get 1–5, I give you $20; if you roll a 6, you give me $1,000. Would you play? Explain your answer.

NOTE: See page 156 for answer sets.

CHAPTER 3

Bill Gates goes back to the diner: Standard deviation and interquartile range

Ok, I know, I can't really see the world's richest man going into a cheap diner in the first place, let alone going back. But I wanted to stick with the example I had, so let's imagine that Bill had enjoyed the good-natured joshing about the design flaws in Windows and didn't have much to do the next day. Also, let's imagine that the diner was much, much busier, with 80 people in total having passed through at some point that morning. Here is a *histogram* showing the salaries of customers.

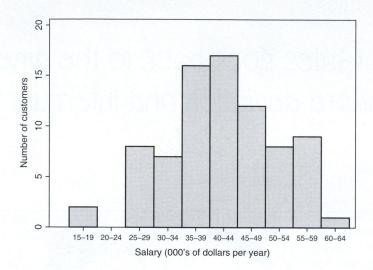

The *x*-axis (going from left to right) shows different salary levels put into different groups (statisticians call these "bins," which has the unfortunate implication that the data are garbage). The *y*-axis (going up and down) shows the number of people within each salary level. For example, the histogram shows that 16 of the individuals who had eaten in the diner have salaries in the range $35,000–$39,000 per year.

You can't work this out from this histogram, but I used the raw data to calculate that the mean salary was $42,360. So we took all those data points and turned them into just a single number. Now, there is an old joke that goes something like: a statistician had his head in the oven and his feet in the fridge. When he was asked how he felt, he said, "On average, pretty good." From this we learn two things: (a) statisticians tell bad jokes (am I repeating myself here?) and (b) a single number often doesn't describe a data set that well. Accordingly, it is generally a good idea to report not just a mean or median—what statisticians call the *central tendency* of the data—but some measure of how much the data vary—what statisticians call a *measure of spread* or a *measure of dispersion*.

One common measure of spread describing how much the study data vary is the *standard deviation*. The standard deviation is calculated from the data using a formula that I won't go into here (roughly speaking, you calculate the difference between each value and the mean, square it, take the mean of all the squares, and then take the square root). The thing to remember is that if the standard deviation for our data set was small, it would mean that everyone has pretty close to the same salary; if the standard deviation was large, it would mean that the salaries of the people in the diner vary widely.

To work out just how much variation we have, we can use some simple rules of thumb. The most well known is that "95% of observations are within about two standard deviations of the mean." This is the same as saying that only 5% of customers have salaries more than two standard deviations from the mean. From the raw data I used to create the bar chart, I worked out a standard deviation of $9,616. We can now work out that 5% of salaries are expected to be either higher than $61,592 (mean $42,360 + standard deviation $9,616 × 2 = $61,592) or lower than $23,128 (mean $42,360 − standard deviation $9,616 × 2 = $23,128). As it happens, one customer has a salary above $61,592 and two are below $23,128 (you can see his from the histogram); this is 3 out of 80, or 3.75%, which is reasonably close to 5%.

It is also true (although this doesn't seem to get mentioned much for some reason) that about two-thirds of observations are within one standard deviation of the mean, and that about half of observations are within two-thirds of a standard deviation from the mean. You can test out these rules of thumb using the histogram: for example, it does look as though about two-thirds of the customers have salaries between about $33,000 (i.e., $42,360 − $9,616) and $52,000 (i.e., $42,360 + $9,616).

That is, of course, until Bill Gates walks into the diner. Now we have a mean salary of about $12 million and a standard deviation of, let's see, $100 million. Clearly it is no longer the case that two-thirds of observations are within a standard deviation of the mean because no one in the diner (other than Bill) has a salary anywhere near $112 million and you can't have a salary of negative $88 million (although this guy I know from college is certainly trying). So the general rules of thumb don't work when the data are skewed away from the bell-shaped curve statisticians call the *normal* distribution (see *Chutes and Ladders and serum hemoglobin levels: Thoughts on the normal distribution*). Here is the normal curve on our data set without Bill Gates. As you can see it isn't a bad fit, which is why our rules of thumb work ok.

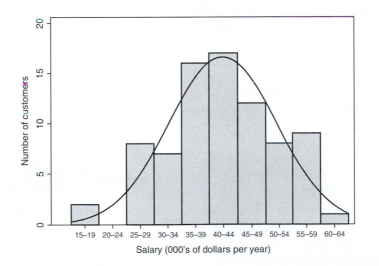

When Bill Gates walks in, however, we have data that we'd describe as "skewed": they don't seem to fit the normal distribution at all. What should you do about standard deviations if you don't have a good fit? Remember that the first time Bill Gates went into the diner, we said that you could use a median instead of a mean as the average. The measure of spread you use with medians is not a standard deviation, but what is called the *interquartile range*. The median is "halfway" along the data; comparably, the *quartiles* are a quarter and three-quarters of the way along. Using the data set in the histograms, I found the median income to be $41,900 and the interquartile range to be $36,000 to $49,300. These three numbers allow you to see a bunch of things immediately. For example:

- 50% of the customers have salaries of more than $41,900 and 50% have salaries of less than $41,900
- 25% of the customers have salaries of more than $49,300
- 25% of customers have salaries of less than $36,000

- 50% of customers have salaries between $36,000 and $49,300
- 25% of customers have salaries between $36,000 and $41,900
- 25% of customers have salaries between $41,900 and $49,300

Bill Gates causes the mean salary and standard deviation to go haywire, but the median and interquartile range remain pretty constant (e.g., the upper quartile goes from $49,300 to $49,500). This is one reason why we said, last time, that if data are very skewed, you often get a fairer reflection of the data if you use medians (and therefore the interquartile range) than means (along with standard deviations).

Here is another reason to use the median and interquartile range. I analyze the results of cancer studies and one of the first things I report in a scientific paper is the general characteristics of the patients in the study: how old were they? What is the ratio of male to female? How many had early stage cancer and how many had advanced disease? Let's imagine that the age of the patients in a study followed very closely to the normal distribution. I could report a mean and standard deviation, knowing that any reader could then work out whatever they wanted about the distribution of ages. But the point is, they are not going to. You can hardly see a busy cancer doctor thinking, "Ok, a mean of 64.3 and a standard deviation of 9.8; half the patients are within two-thirds of a standard deviation of the mean, that is, $64.3 + 9.8 \times 0.667$, which is—wait a minute, where's my calculator?" You can just glance at the median and interquartile range and get a good, quick idea about the sort of data that you are dealing with.

In other words, the median and interquartile range are very useful for describing a data set. And this is exactly what we want them to do: everything I have been talking about here—means, medians, standard deviations, interquartile ranges—are known as *descriptive statistics*.

• Things to Remember •

1. Means and medians are useful for describing a data set. Means and medians are types of average, or central tendency.
2. You generally want to know not only the average of a data set, but how much the data vary around that average: a measure of spread.
3. The measure of spread normally reported with a mean is the standard deviation.
4. The measure of spread normally reported with a median is the interquartile range.
5. If data follow something close to a normal distribution, the mean and standard deviation can be used to work out all sorts of things about your data, but you have to do some calculations.
6. The median and interquartile range give quick information about data without the need for any calculation.
7. The median and interquartile range are also useful for describing data that are skewed.
8. Statistics used to describe a data set, means and medians, standard deviations and interquartile ranges, are known as descriptive statistics.

for Discussion

· ·

1. The upper and lower quartile are sometimes described as the 75th and 25th *centile* (or *percentile*). Explain this.

2. When talking about the interquartile range I said things like 25% of the customers in the diner have salaries of "$49,300 or more." When talking about standard deviations, I said 5% had salaries above $61,592 or below $23,128. Why did I sometimes say "*x* or more" and sometimes "higher than *x*?"

3. Is it really true that 95% of observations are within two standard deviations of the mean, even for a perfectly normal distribution?

NOTE: See page 157 for answer sets.

CHAPTER 4

A skewed shot, a biased referee

Like all great moments in my life, I remember it as if it were yesterday. England was playing Spain in the European soccer championships, a Spanish player mis hit a shot, which then fell to a team mate, who slotted it past the English keeper for a goal. But then the referee disallowed the goal for offside (relief!), even though the replay showed that the goal ought to have counted.

I couldn't believe it: *there had been a bad decision on the soccer field and, as an England fan, I wasn't going to suffer as a result.* The British press the next day were full of praise for "our brave lads" (England had squeaked through on penalties); the Spanish press were outraged at what had obviously been a biased referee. In my view, if the Spanish player hadn't skewed his shot in the first place, we wouldn't even be talking about the referee. But then again, you'd hardly call me unbiased about it.

Skew means "off to one side." Statistics can also get "off to one side," for one of two reasons. First, sometimes that is just how the data are: you have more observations on one side than the other. The following graph is from a study of US adults and gives the body mass index (which is weight in kilograms divided by the square of height in meters):

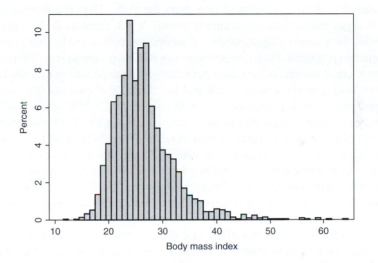

This data set is skewed to the right. There are some Americans who are perhaps a little underweight, but the Americans who are on the heavy side are often very heavy (20—25 is considered "normal"). These data are right-skewed because observations above the median tend to be further from the median than observations below the median. As a result, the mean is higher than the median (26.5 vs. 25.7).

Left-skewed data is when the mean is less than the median. Here is an interesting example of left-skewed data—duration of pregnancy:

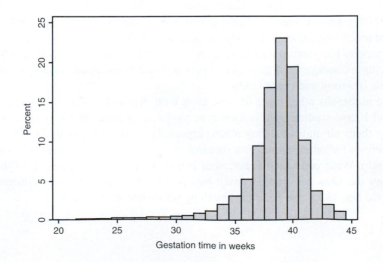

There is a long tail off to the left of premature infants, but not a tail off to the right of infants born much later than their due date. This is because a very long pregnancy can be dangerous, and doctors don't let any woman carry a baby for more than two weeks longer than normal (although some women appear to have slipped through the net). As a result, the mean is lower than the median.

Another meaning of skew is "skewed away from the truth." Here is a famous example of an opinion poll that got things quite spectacularly wrong. When Franklin D. Roosevelt ran for re-election in 1936, the *Literary Digest* conducted an opinion poll to predict the result on election day. There were two problems. First, they selected much of their sample from the telephone book. Only a relatively small number of wealthier Americans had telephones during the Great Depression and richer folks tended not to like FDR and his "New Deal" policies. The second problem was that the poll had a very low response rate, with only about 20–25% of those polled returning their postal ballot. It seems likely that those voters who didn't like FDR were especially motivated to give the *Literary Digest* a piece of their mind. You'd probably want to say something like "the *Literary Digest* used a skewed sample of voters." As it happens, statisticians tend to reserve the word "skew" to describe data that is off to one side or another. In fact, you can work out the "skewness" of a set of data using a formula just as you can work out the mean and standard deviation. To describe an error in experimental methods or statistical analysis that leads to an incorrect estimate, statisticians use the word *bias*.

The *Literary Digest* study was biased because those responding to the survey were not a representative sample of American voters. A medical study might also be biased by this sort of *selection* bias. For example, a study might examine the survival rates of heart attack patients, comparing those undergoing a heart operation with those not treated surgically. This study would be biased because some patients are too sick to go through an operation; only the healthier patients go for surgery. So you would expect survival rates to be better in surgical patients even if the surgery didn't help at all.

You can also get bias even if you select your sample carefully and fairly. For example, if you are asking people questions as part of your study, the way that you ask them can introduce bias. An obvious example is "push polling," where political campaigns conduct phony opinion polls: "If you found out that Brown, candidate for governor, had fathered four children out of marriage and had paid off a judge to escape a bribery charge, would that make you more or less likely to vote for him?" My favorite case of biased questioning came in a study suggesting that rates of adultery were much lower than previously thought, only 2–3% rather than 15–20%. It turned out that the researchers had interviewed married couples sitting together in their own home. This is hardly likely to encourage frank answers to personal questions about something which almost everyone feels is wrong and tries to hide.

There are numerous other types of bias, each with their own name (if you are interested, a colleague and I have studied what is known as *verification bias*). But it is probably not worth remembering them all and what they mean, especially as statisticians themselves disagree on what to call things (when explaining our research at a conference, someone said, "Oh, you mean detection bias."). What you need to remember is that skew is out there as part of the world (the Spanish player did skew his shot on goal); bias is what we can sometimes introduce when we study the world but, like a referee, it is something we should try to avoid.

• Things to Remember •

1. Skewness describes the distribution of data.

2. Data are skewed if there are more observations below the mean than above, or vice versa.

3. The greater the proportion of observations above or below the mean, the more skewness you have.

4. Bias describes a problem with the design, conduct or analysis of a study.

5. A study is biased if the methods or statistical analyses cause an estimate to be too high or too low.

6. My joy was short lived. England lost the next time they played—a game they should have won.

SEE ALSO: *If the normal distribution is so normal, how come my data never are?*

for Discussion

1. How would you avoid selection bias in the surgery study?

2. Imagine that you were conducting a study on cheating at college. Like the adultery research, this involves questions about bad behavior. How would you encourage truthful answers?

NOTE: See page 158 for answer sets.

CHAPTER 5

You can't have 2.6 children:
On different types of data

My mother has a statistical insight

Growing up, my mother used to tell me, "Statisticians have got it wrong: you can't have 2.6 children." Now before any amateur psychologists out there chime in to tell me that I obviously became a statistician to annoy my mother, let me say that: (a) she is completely correct and (b) she has hit upon something quite profound. My mother's comment came after a survey finding that the "average" British woman had 2.6 children. This "average" was clearly what statisticians call a mean. To get a mean, you add everything up and divide by the number of observations, so you'd get a mean of 2.6 if there were, say, 1 million women and a total of 2.6 million children.

The other type of average most often used by statisticians is the median, the number higher than half the observations. My guess would be that the median number of children per woman in the survey was 2, that is, 50% of women have 0, 1 or 2 children, and 50% have 2, 3 or 4 (or some number even less conducive to a peaceful Sunday morning).

From my mother's point of view, the big difference between the mean and the median is that one is an artificial abstraction—a mathematical calculation—and the other relates to something you can actually go out and see. Generally speaking, someone in a data set will have a value at the median (a family of 2 kids); that often isn't the case for the mean (you can't have 2.6 children). The same is true of the measures of spread generally reported alongside means (standard deviation) and medians (interquartile range).

Why not stick to the median?

So why use "artificial" numbers like the mean or standard deviation and risk giving my mother something to complain about? The quick answer is that you can use the artificial numbers to answer a whole host of questions; if you want to use the real data, you have to look it up each time. Let's imagine that we had a data set consisting of heights from a sample of 10-year-old boys. A well known rule of thumb is that "95% of observations are within two standard deviations of the mean." But you can also calculate that about two-thirds of observations are within one standard deviation of the mean, that 90% of observations are greater than the mean minus 1.28 standard deviations and, just to show that you can do pretty much whatever you want, it is also true that 86.4% of observations are less than the mean plus 1.1 standard deviations. So, give me a mean and a standard deviation and I can immediately answer questions such as: What height is exceeded by only 5% of boys? What proportion of boys are more than 5 feet tall? What are the heights between which 50% of the boys' heights can be found? There is no need to go back to the data set and look anything up.

I can also do some hypothesis testing. If I have the mean and standard deviations of boys raised as vegans, and similar data from a control group from the general population, I can work out whether avoiding animal products affects growth in boys.

How to use 2.6

As you know, mothers are usually right, and my mother was certainly right in this particular case—citing an average of 2.6 children per woman is a bit of a silly statistic. We use statistics to do one of two things: *estimate* or *infer*. Take, for instance, the data on the height of 10-year-old boys. Estimates for the average height (e.g., a mean) and how it varies (e.g., a standard deviation) are great for description. We could look at these statistics and get a great idea about the usual heights of 10-year-old boys and just how common it is to be more than, say, 6 inches taller or smaller than average. We might then use our statistics to do something useful, like design a playground, or choose a range of sizes for a line of rain coats.

Inference—hypothesis testing—also has a practical application. Say we conduct a statistical hypothesis test to compare the heights of boys raised on vegan and regular diets, and conclude that the vegan diet does indeed appear to inhibit growth. We might then consider advising parents about the implications of raising their children on restrictive diets.

It is not clear how we could use "an average of 2.6 children" for testing a hypothesis. For example, imagine that we had a data set of women, giving the number of children they had given birth to and where they lived. We can't use the mean number of children to test a hypothesis such as "women in rural areas have more children than those living in cities." This is because statistical tests that use means assume that data can take a lot of different values, what statisticians call a *continuous* (or *quantitative*) variable. Height is a good example of a continuous variable: a 10-year old boy can be 4 ft 2 in., or 5 ft 2 in. or 4 ft $7\frac{1}{2}$ or pretty much anywhere in between. Family size is more like what is called a *categorical* variable, because almost all families (in industrialized nations, at least) have a family size in one of a limited number of categories: zero, one, two, three or four children, and so on. Statisticians don't like to use the sort of statistical tests designed for continuous variables on categorical variables like family size.

Moreover, "an average of 2.6 children" doesn't even give us a clear idea of how many children women typically bear. It sounds as though most women have either 2 or 3 children, but that might not be it at all; it might be that most women have either 1 or 2 children, and a few women have lots and lots. This is an example of the general rule that a single number can rarely be used to describe data and more specifically, that measures of spread should be reported alongside estimates. The problem is that the measure of spread associated with the mean is the standard deviation, and this means very little if data are skewed (see *Bill Gates goes back to the diner: Standard deviation and interquartile range*). For example, here is a histogram with some example data for the number of children per woman:

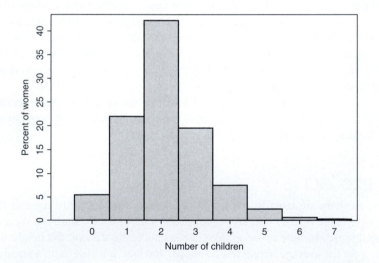

I couldn't actually find a data set with a mean of 2.6 (I think the survey my mother was referring to was conducted in 1968 or something). This data set has a mean of 2.1 and a standard deviation of 1.1. Applying the rule that 95% of the data are within 2 standard deviations of the mean, you get that 95% of families have between −0.1 and 4.3 children—having a negative number of children is even more ridiculous than having 2.6. So, what statistics should we use in place of our mean of 2.1? The median (2) and interquartile range (1 to 3) aren't that helpful. For example, you know that 25% of families have 1 or fewer children, but (a) I would have kind of guessed that and (b) this would be true if either 1% or 24% of families were childless.

If not means or medians, then we are left with the histogram. And that's fine; it tells you pretty much everything you need to know. Or you could have a table like this, from which you could deduce not only that, for example, 7.4% of women have 4 children, but that 96.7% have 4 or fewer and 3.3% have more than 4.

Number of children in the family	Percentage	Cumulative percentage
0	5.5	5.5
1	22.0	27.5
2	42.2	69.7
3	19.6	89.3
4	7.4	96.7
5	2.4	99.1
6	0.6	99.7
7	0.3	100

True, we have lost the sound bite, with neither the table nor the graph as punchy as "an average of 2.6 children." And neither is likely to annoy my mother, which I see as a considerable downside. But here, finally, is something you could actually use.

• Things to Remember •

1. Statisticians sometimes calculate numbers from a data set, the mean and the standard deviation being good examples.

2. These numbers often take values that don't occur on the data set.

3. Statisticians sometimes choose particular numbers from the data as being illustrative, such as the median and quartiles.

4. Means, medians, standard deviations and interquartile ranges are used to describe variables that can take a large number of different values. These are known as continuous or quantitative variables.

5. Variables are sometimes best described in terms of categories: sometimes means and medians aren't used and the statistician just gives the number and percentage in each category.

SEE ALSO: *Numbers that mean something: Linking math and science*

for Discussion

....

1. Does the median (or, say, upper quartile) always take a number that is part of the data set?

2. I described a continuous variable as one that can take "a lot of different values." How many different values is "a lot?"

3. Here are some variables. Which of these are continuous and which are categorical?

   a. Height

   b. Gender

   c. Years of education

   d. Pain score

   e. Depression

   f. Income

   g. Race

   h. Unemployment rate

4. Saying that an "average of 2.6 children is a silly statistic" allowed me to make some nice teaching points about different types of data. But as it happens, the "average" number of children that a woman bears over the course of her lifetime is actually pretty useful. How do you think that this statistic is used?

NOTE: See page 159 for answer sets.

CHAPTER 6

Why your high school math teacher was right: How to draw a graph

Another difference between normal people and statisticians: a normal person might say, "High school . . . those were the happiest days of my life." Statisticians tend to be more particular in stating that it was "high school math" that made them happiest. As it happens, much of the math I did in high school is far more advanced than what I do in my day to day work. (I don't need calculus on the average Tuesday morning.) But one thing I learned really stuck with me—how to draw a graph.

What I was taught about graphs was that you have an x-axis and a y-axis and, to draw a line, you state the value of y in terms of x, for example, $y = 5x - 4$. I was also taught that you can also put in other powers of x (e.g., $y = 1.4x^2 - 5x - 4$). Doing so allows you to have a curve, rather than a straight line (this is called a "non-linear" relationship). The number of times this curve can change direction is related to the number of different x terms that you have. From left to right, here are graphs for $y = 5x - 4$, $y = 1.4x^2 - 5x - 4$, $y = 0.5x^3 + 1.4x^2 - 5x + 4$.

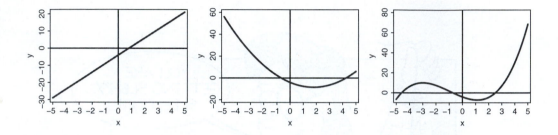

What I also did in school was to have the x- and y-axis represent something, like shoe size and height, and put a mark for each observation at the appropriate coordinate: for someone 5 ft $11\frac{1}{2}$ in. with a shoe size of $10\frac{1}{2}$ (which may or not be me, I couldn't possibly say), I'd draw an imaginary line up from the x-axis at $10\frac{1}{2}$ and an imaginary line across from the y-axis at 5 ft $11\frac{1}{2}$ in. and then put a dot where the two lines meet. I would then draw a line through the graph so that the line comes closest to the dots representing each person's height and shoe size.

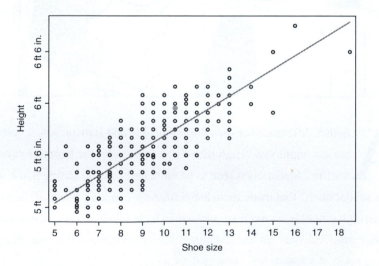

It turns out that this is just about all you need to know for most graphs. This begs the question of why good graphs are so few and far between in the scientific literature.

Here is a graph that is fairly typical of what you see in reports and papers:

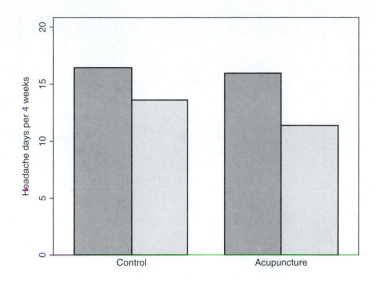

This shows the results of a clinical trial of acupuncture for headache. The bars show the number of days with headache per month before treatment (dark gray) and after treatment (light gray). Or how about this, from a survey on the state lottery:

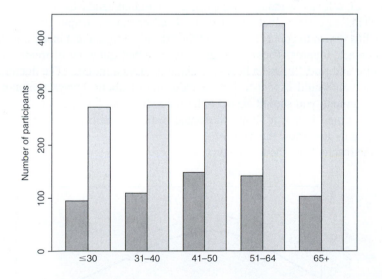

This shows the number of survey respondents who played (dark gray bar) or did not play (light gray bar) the lottery in the previous 12 months, separately for different ages.

Now, for those of you still awake, a short review of why those graphs are so bad: (a) they are boring to look at, (b) they don't give an immediate visual impression of the results and (c) they don't provide information that anyone could actually use.

So, back to high school: let's present the results on an *x* and *y* graph. Here are the results of the acupuncture trial:

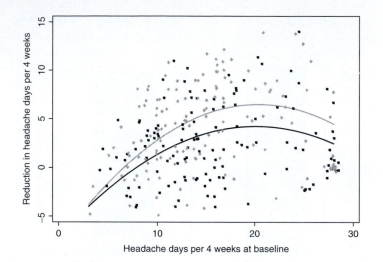

The *x*-axis shows patients' level of headache at the start of the trial; the *y* axis shows whether they got better (a reduction of "+5" means that the patient had 5 fewer days with a headache at the end of the trial than they had at the beginning). The gray diamonds show the results of each patient in the acupuncture group; the black squares show the results of the patients in the control group. In general, it looks as though, irrespective of baseline headache, the gray dots are higher up the graph. This suggests a greater reduction in headache in the acupuncture group. I have also shown a gray line drawn to come closest to all the gray diamonds and a black line drawn to come closest to the black squares. These lines give the expected outcome of treatment: if a patient went to a doctor and said, "I have a headache about 20 days a month." The doctor could say, "A year from now you would expect to have a reduction of about 3 days a month. If you have acupuncture treatment, you should expect to have a reduction of 5 days a month." Another nice thing about this graph is that it shows the actual results of each patient in the trial. It's what graphs are meant to do—show the data.

Here are the results from the lottery survey:

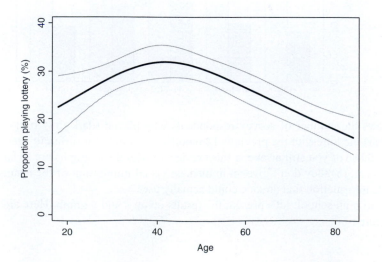

The black line shows the proportion of survey participants playing the lottery in terms of age. The thin grey lines show what is known as the 95% confidence interval, which reflect a plausible range for the relationship between age and lottery playing (see *How to avoid a rainy wedding: Variation and confidence*). You can instantly see that young and old play the lottery less than those of middle age, with a peak around age 42. The nice thing about this graph is that it gives an immediate visual impression of the data. The graph might, for example, be of interest to a psychologist trying to understand gambling addiction.

A final point: have a look at the far left of the acupuncture graph. It looks as though acupuncture actually makes things worse for patients who didn't have many headaches at baseline. But look again and you'll see that this is based on only a small number of patients. This is a reminder that we have to be careful about applying average statistical results to patients at the extremes. Then again, at least you have something to apply other than a boring gray bar.

• Things to Remember •

1. Often in research, we want to understand one thing (such as playing the lottery) in terms of another (such as age).

2. The thing we want to understand can be called y.

3. The thing we use to understand y is called x. So y might be lottery playing and x might be age.

4. Drawing a graph of y and x, just like you did in high school, is a good way of understanding the relationship between them.

5. You can mark a point at the x and y of each observation (this is called a *scatterplot*).

6. You can also draw a line or curve closest to each point.

7. You can use different colors for your points and lines to indicate different categories (such as headache separately for patients who did and did not receive acupuncture).

for Discussion

1. Can you always draw a line?

NOTE: See page 162 for answer sets.

Chutes and Ladders and serum hemoglobin levels: Thoughts on the normal distribution

One of the great challenges of parenthood is how to lose games of chance. How can I let my son win at Chutes and Ladders* (thereby improving his self-esteem and decreasing family tension) without cheating (which, as I understand it, would send the wrong message)? I can't, of course, but "just one more game" does at least allow me to reflect on the nature of statistical distributions.

---

*Chutes and Ladders is a trademark of Hasbro.

Chutes and Ladders is a bit like a coin flip, in that there is exactly a 50:50 chance that I'll win a game. So if you tell me that we are going to play a certain number of games, I can tell you the probability of each possible combination of wins and losses. As an easy example, if I play two games of Chutes and Ladders with my son, there is a 25% chance I'll lose both, a 50% chance that we'll each win one and a 25% chance that he'll throw a hissy fit. I can show this as a bar chart. The *y*-axis gives the probability that I'll win each particular number of games shown on the *x*-axis:

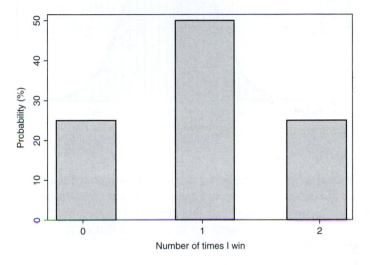

The math is a bit more complicated for four games but, as it turns out, there is a 37.5% chance that we split it with two games each and a 6.25% chance of a total meltdown.

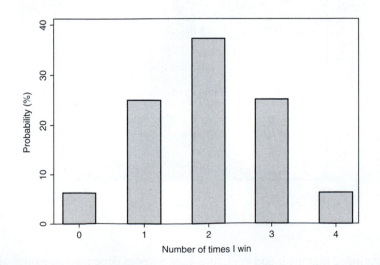

Something you might notice here is that this second graph is starting to look a little bit like the bell-shaped curve of the normal distribution. Now let's imagine a *really* wet weekend in which we play 100 games of Chutes and Ladders:

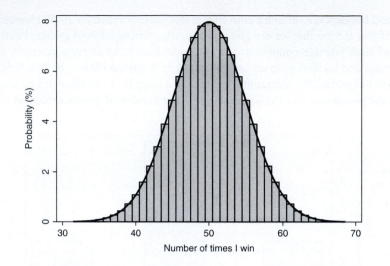

We now have something that is really very close to a normal distribution (the normal curve is plotted over the histogram so you can see just how close it is). We also have something that looks very much like many natural biological phenomena. As an example, the following graph is the distribution of hemoglobin (a substance in the blood that carries oxygen) in a cohort of Swedish men aged 35–50:

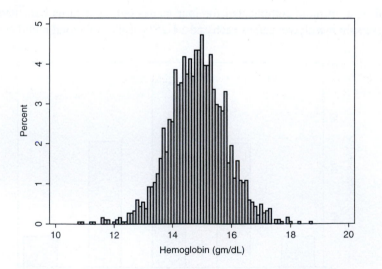

If you concluded that the blood of middle-aged Swedes depended on games of Chutes and Ladders, you wouldn't be far wrong. Like the outcome of a dice-throwing game, a man's hemoglobin level is the result of numerous chance events—genes, environment, diet, lifestyle and medical history—all added together. For example, imagine we look at just four things that affect hemoglobin:

Event	Effect on hemoglobin
Has a gene for producing a lot of hemoglobin	Increase
Just returned from a trip to the mountains	Increase
Has been eating a poor diet lately	Decrease
Recent illness	Decrease

If we assumed, for the sake of argument, that everyone has a 50:50 chance of each event and that each leads to an increase or decrease by about the same amount, then a histogram of hemoglobin levels would look very similar to the one showing the results of four games of Chutes and Ladders. There are hundreds of influences on hemoglobin, not just these four and, as we saw in the Chutes and Ladders data, when you add up a lot of chance events you get a normal distribution. To a statistician, the normal distribution is a complicated formula including e, μ, π and σ all raised to the power of each other. But the formula for the normal distribution is just a mathematical way of describing what you get when you sum up a large number of chance events.

One set of chance events that is of particular interest to statisticians is the results of experiments. As an example, we'll use a psychology experiment investigating influences on IQ test scores in African Americans. In the experiment, groups of African American students take the same exact IQ test. Half are told that they are receiving a test of innate intelligence. For the other half, however, IQ isn't mentioned at all—participants are just told that they are taking part in an experiment to evaluate differences in problem-solving styles. Let's say that, at the end of the study, 15 of the 50 (30%) students in the "innate intelligence" group score at or above the national US average compared to 25 of 50 (50%) of those in the "problem-solving" group—a difference of 20%. This finding might be taken as evidence that apparently poorer IQ scores in certain racial groups result from test-related anxiety.

Now if I repeated the experiment, we wouldn't expect to get exactly the same result—we'd expect some chance variation. For example, if the pass rates the second time around were 32% and 48%, you'd probably say that I'd done a pretty good replication. The following graph shows the results of the study if I'd repeated it 100,000 times and there was really no difference in scores depending on what students were told about the test:

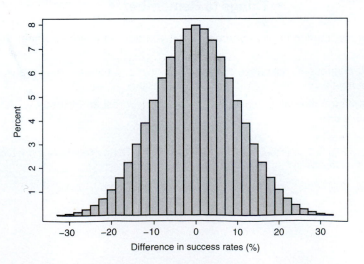

We get the normal distribution for the results of the IQ experiment because we are adding up the result of a lot of chance events. Every student in the study has a certain chance of scoring above average on the test, just as they have a certain chance of winning a game of Chutes and Ladders. The number scoring above average, just like the number of times I win at Chutes and Ladders, follows the normal distribution.

I drew the last graph to show the sort of results you would get if there was no effect on test scores of telling students different things about the test. As you'd expect, the most common result is no difference between groups. But the graph shows that sometimes you get more people scoring below average in the "problem-solving" group, just as sometimes my son will win more games of Chutes and Ladders than I will. The shape of the graph is pretty close to the graph of hemoglobin levels because, like hemoglobin (and the results of multiple games of Chutes and Ladders), what we observe is the sum total of a large number of random events. By the same token, you would also see a normal distribution if you plotted the results of multiple studies calculating a mean—say, studies of the mean hair length of guys at parties (see *Long hair: a standard error of the older male*).

The normal distribution—the sum of a lot of random events—can therefore be used to describe both the natural variation that we observe in the world around us and the hypothetical variation of research results. Moreover, the normal distribution applies whether our result is a difference, such as in test scores between individuals taking different types of IQ test, or whether we are making an estimate, such as mean male hair length. This means we can use the mathematical formula for the normal distribution both to help describe data sets and to work out whether our results are interesting. As you can see from the graph, a 20% difference between groups—the data actually observed in the study—would be unusual if test performance was unaffected by whether students were told they were taking part in an IQ test versus a problem-solving experiment. Accordingly, we can be pretty confident that how IQ tests are presented does indeed affect the test scores of African American students.

• Things to Remember •

1. The normal distribution is what you get when you add up a large number of random events.

2. The normal distribution describes the variation of many natural phenomena, such as hemoglobin levels.

3. The normal distribution also describes the variation in results of a study, were we to repeat it many times.

SEE ALSO: *Long hair: A standard error of the older male; The probability of a dry toothbrush: What is a p-value anyway?*

for Discussion

1. Is the distribution of the results of a game of chance, such as Chutes and Ladders, really a normal distribution?

2. Why doesn't the graph of hemoglobin levels in the Swedish men follow a perfectly smooth curve?

NOTE: See page 163 for answer sets.

If the normal distribution is so normal, how come my data never are?

One of the first data sets I looked at when I was first learning statistics had a number of missing observations. I was told that this was totally normal. I also noticed that the data followed the bell-shaped curve of the normal distribution. This, I was told, was not normal at all. One of my lecturers became rather excited, commenting that, "They say it never happens, but look—here is an example. Just goes to show that you *can* get a normal curve." Now, I think what they were trying to tell me was that it wasn't normal to get normal data. Indeed, non-normality seemed to be the norm. But I couldn't be sure.

With the benefit of hindsight, I can see that what was behind my lecturer's comment about normal distributions being abnormal was that he was a biostatistician who studied medical data. Normal distributions are quite common in nature but not so much in medical research. As an example, here are some data I have been looking at recently. The following graph shows the distribution of prostate specific antigen (PSA) levels in men undergoing surgery for prostate cancer. In simple terms, PSA levels roughly correlate with the number of cancer cells. So the higher your PSA, the more cancer you have (the correlation is only approximate because increases in PSA can be caused by diseases other than cancer).

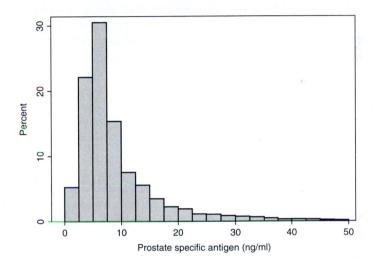

And, just for the sake of it, here are data from a totally different area of medicine. These are pain scores from patients with migraine headache:

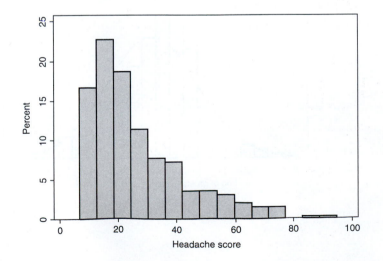

Both graphs look pretty similar to each other and pretty dissimilar to a normal distribution. The simple explanation for what is going on here is that medical research typically involves studying patients with some kind of disease. By definition, these populations are not normal; they have presented for treatment exactly because they have something wrong. Perhaps this is what was behind my professor's comment about the rarity of normal distributions in medicine— you hardly ever see normal distributions in medicine because you hardly ever study the "normal" population as a whole, only unusual subsets.

A more mathematical way of saying this is that whereas normal processes usually involve addition (see *Chutes and Ladders and serum hemoglobin levels: Thoughts on the normal distribution*), disease processes often involve multiplication. Cancer is a good example. Cancer cells divide and grow and tumors therefore double in size every few months. In the case of headache, a series of severe headaches leads to a number of changes—such as increases in anxiety and muscle tension, or overuse of analgesics—that increase the risk, and severity, of subsequent headaches. For example, I'll have one mild headache, you'll have a severe headache and then another milder one as a result, so your headaches are exponentially worse than mine.

If you want to convert a multiplication into an addition, you use logarithms. Take $10 \times 100 = 1000$; $\log(10) = 1$, $\log(100) = 2$ and $\log(1000) = 3$, so $\log(10) + \log(100) = \log(1000)$.

Let's calculate the log of our headache and PSA data and see what we get. We'll do what statisticians usually do and take the "natural" log using the constant known as e for each person's headache score or PSA level. I've shown the curve for the normal distribution on each graph so you can tell whether or not the data are close to being normally distributed.

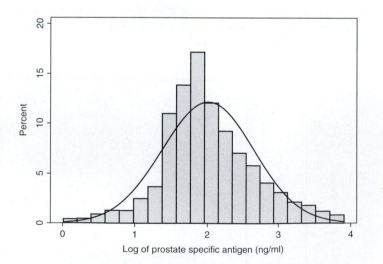

Log of prostate specific antigen (ng/ml)

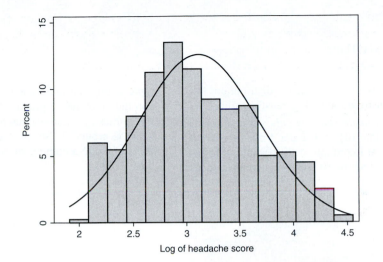

These data look like a pretty good approximation to the normal distribution. From this I would conclude that the *rate* of cancer growth is normally distributed in patients undergoing surgery and that there is some normally-distributed tendency to headache in patients with headache disorders.

This illustrates a key aspect of statistics: statistics is about linking math to science. I am a biostatistician involved in medical research, so what I do is link math to biology and medicine; economic analysis is about linking math to human economic behavior; good use of statistics in psychology links math to the human psyche. You might hear it said that the purpose of log transformation is to "bring down high values" or to "allow the use of parametric statistics." But that is looking at numbers in reference only to other numbers. We use log transformation when we believe the underlying process involves multiplication—the growth of cancer being an obvious example.

• Things to Remember •

1. Despite being called "normal," you rarely see a normal distribution in some areas of statistics.
2. Normal distributions are rare when the data come not from the whole population, but from a special sample, such as medical patients.
3. The normal distribution results from the addition of numerous random events.
4. Many phenomena result from the multiplication of random events.
5. Logarithms change multiplication into addition.
6. Non-normal data can sometimes be converted to a normal distribution by using the logarithm of the data.

SEE ALSO: *Chutes and Ladders and serum hemoglobin levels: Thoughts on the normal distribution; Numbers that mean something: Linking math to science*

for Discussion

· ·

1. Can you transform all skewed distributions to a normal distribution by log transformation?

2. *For enthusiastic students only:* At one point I said that $\log(10) = 1$. Later, I mentioned e. If you look at the graph of PSA values, you can see that a PSA of 10 comes just after the peak representing the most common PSA level. If you then look at the graph of log transformed PSA values, you can see that the most common value is around 2. So my log transformation turned 10 into a number slightly over 2, rather than 1. Why?

_____

NOTE: See page 163 for answer sets.

CHAPTER 9

But I like that sweater: What amount of fit is a "good enough" fit?

Sometime each fall, we bring up the winter clothes from the basement for our children to try on. Most clothes are immediately thrown into one of three piles: "comically small," "just about right" and, for hand-me-downs, "still too big." This leaves us with a pile of clothes that either we think fit and the children don't ("Those yucky gloves are too small," says my daughter) or the children want to wear despite suffering something close to compression injury ("But I *like* that sweater," says my son). This all goes to show that, unless it is ridiculously obvious, whether something fits or not is a judgment call.

One of the first things anyone learns about statistics is how the choice of mean or median depends, at least in part, on whether the data follow the normal distribution. (Remember what happens to the mean salary when Bill Gates walks into a diner? See *So Bill Gates walks into a diner.*) Later on, when you come to hypothesis testing (see *Choosing a route to cycle home: What p-values do for us* and the chapters that follow), you might learn that a choice between different tests, such as the *t* test or Wilcoxon also depend on how the data are distributed. Textbooks often describe statistical analysis as a sort of two-step process: have a look at the data, test whether or not it is close to a certain statistical distribution (such as the normal distribution), and then decide how to analyze it (e.g., if it is normal, use the *t* test, otherwise use the Wilcoxon). But you rarely see this sort of analysis described by statisticians.

Statisticians don't typically seem to worry too much about whether or not the data are a close fit to the normal distribution because they realize that statistics isn't football, and no one is going to throw a flag and send you back 10 yards if you are caught breaking the rules. In fact, there aren't really many "rules" at all.

Here are some data that clearly don't fit the normal distribution. These data are typical for tests of knowledge (e.g., about computers) or physical ability (e.g., time to run a mile) converted to a 0–100 scale. There is a peak at zero because a bunch of people know absolutely nothing about computers (e.g., they have never used them) or have some kind of disability that prevents them from completing the run. Then there are more very high achievers than you'd expect by chance because some people choose to study English literature, or train as athletes, enhancing their natural ability or knowledge. Since the data are non-normal, you can't use any statistics that assume the normal distribution—right?

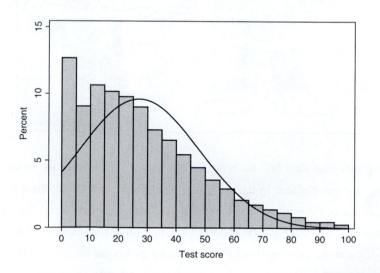

The table below compares estimates calculated using the mean (27) and standard deviation (20.7) with those observed in the data set. For example, in a normal distribution the median and the mean are the same, so we'd expect the median to be 27. However, it is 23, which isn't too far off. Most of the other estimates are pretty close and some are absolutely spot on—exactly 95% of the observations are within 1.96 standard deviations of the mean. Using means and standard deviations only breaks down at the extremes. You can't score less than zero and, as you can see from the histogram, there are more very high performers than expected.

Estimate	Expected from the normal distribution	Observed in the data set
Median	27	23
Interquartile range	13 to 41	11 to 39
Proportion within 1.96 standard deviations of the mean	95%	95%
Proportion within 1 standard deviation of the mean	68%	67%
Proportion within two-thirds of a standard deviation of the mean	50%	46%
Proportion higher than 2.33 standard deviations greater than the mean	1%	3%
Proportion lower than 1.64 standard deviations below the mean	5%	0%

Clearly, sometimes the data are so skewed away from normal that using the normal approximation is clearly unsound (e.g., when Bill Gates went into the diner, the mean salary became $125m). And not everyone would agree that using the mean of 27 instead of the median of 23 is "good enough." (What if 25 was the passing grade?) As a result, deciding whether the data fit a distribution is perhaps as much of a judgment call as deciding whether last year's winter jacket fits a growing child.

• Things to Remember •

1. Many statistical procedures are based on the assumption that the data are normally distributed.

2. These procedures include the use of the mean to describe a data set and hypothesis tests such as the *t* test, but there are many others.

3. There are no clear rules for determining whether a data set is close enough to the normal distribution to make it reasonable to use statistical procedures that assume normality. Ultimately, it is a judgment call on whether the method provides a "good enough" approximation.

∴• **SEE ALSO:** *If the normal distribution is so normal, how come my data never are?*

for Discussion

1. Isn't statistics meant to be very precise? Don't we prefer "28.29%" to "about one in three?"

2. The histogram showing test scores is skewed to the right. Why would that be?

3. It is generally said that "95% of observations are within two standard deviations of the mean." To calculate where 95% of the observations were for the table, I multiplied the standard deviation by 1.96 rather than 2. How come?

NOTE: See page 166 for answer sets.

CHAPTER 10

Long hair: A standard error of the older male

My hair is longer than that of my statistician colleagues. This was never much of an issue for me until someone suggested that I was having a mid-life crisis and asked when I would be buying the sports car. So I thought about it for a bit and came up with the idea that the problem was with the other statisticians, not with me: *my* hair was of average length, it was *their* hair that was short. As it happened, I was at a party when I had this flash of inspiration, and this provided me with an immediate opportunity for data collection. Twenty minutes after locating some scissors in the kitchen, the host had asked me to leave, but in the meantime, I did have hair samples from 46 men. The results are shown in the following histogram. Hair lengths are given to the nearest centimeter, so, for example, 9 men at the party had hair less than 2 cm long.

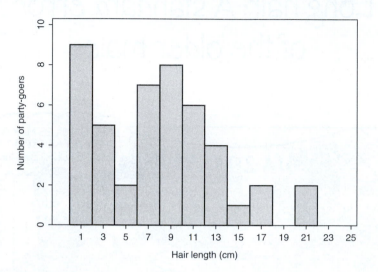

The mean of this data set is 7.5 cm and the standard deviation 5.34 cm. Federal privacy regulations prevent me from identifying myself on this histogram, but I can tell you that my hair was longer than most of the other guys at the party. I decided that I might have been unlucky and resolved never to leave home without a pair of a scissors and measuring tape. When I got data from the next party I found a mean of 9.1, slightly higher than at the first party. This isn't unusual; we don't expect to get exactly the same results every time we run a study.

The results I got from the next five parties are shown in the table. You can think of this as a new data set except that instead of each observation corresponding to a person, each observation corresponds to a study.

Party	Number of men at party	Mean male hair length (cm)	Standard deviation
1	50	9.1	6.85
2	59	8.5	6.78
3	68	7.0	5.81
4	35	8.9	7.97
5	46	7.8	6.53

When I plot the means I get the histogram on the far left. The histogram in the middle shows the mean hair length recorded at 50 parties and the one on the right shows the results of a hard year of 500 parties. (Ok, I admit it. I didn't go to 500 parties, I simulated all this on a computer.)

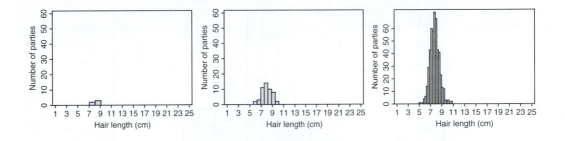

So if I repeat a study (e.g., on male hair) many times, I don't always get exactly the same results (e.g., mean hair length); the results vary following a distribution that is typically pretty close to the normal distribution (see *Chutes and Ladders and serum hemoglobin levels: Thoughts on the normal distribution*). The mean and standard deviation of this distribution—the distribution of our result, mean hair length, from numerous studies—can be calculated. The mean (7.8 cm) is pretty typical of the results in the table, but the standard deviation (0.82 cm) is quite a bit smaller. You can see this by comparing the figure on page 42 with the figure at the top right of page 43. This makes sense—you may well find one guy with a buzz cut or a pony tail at a party, but it would be odd if you went to a party (in New York anyway) and found that everyone was bald or, alternatively, looked like a rock star from the 1970's. (If you don't know what I mean, just Google "Frampton comes alive.")

The histograms above show variation, but they are two very different types of variation. One is *natural* variation that you can actually see. Go to a party and you can directly observe the length of everyone's hair—the guy with the hipster glasses has gone for something short-ish, the jock's head is practically *shining* and the biker dude could definitely do with a trim. The other is *theoretical* variation of study means that you'd see were you to repeat a study a large number of times. Of course, this would be a pretty silly thing to do—I am currently working on a cancer study that included over 22,000 men followed for about 30 years, and I am hardly going to run that study multiple times just to see what happens.

The point about the theoretical variation is that you can calculate what it would be using formulas developed by statisticians. The statistic that is used to describe this variation is the standard error. The formula for the standard error of a mean is the standard deviation divided by the square root of the sample size. The standard deviation of hair length at the first party was 5.34 cm. When you divide this by the square root of the number of guys at the party (46), you get 0.79 cm. This is around what we got for the standard deviation of the means of the repeated studies reported above (0.82 cm). So, a standard error is the same as the standard deviation of the results (e.g., means) of repeated studies.

There are also formulas for the standard error of other statistics, such as proportions. For example, if we had asked the 46 guys at the party whether they had cut their hair in the last 6 weeks and 23 said that they had, we can calculate the standard error for 50% of 46 as 0.074. When I simulated 500 parties, I got the following figure, which has a standard deviation of 0.070—pretty close to what we expect.

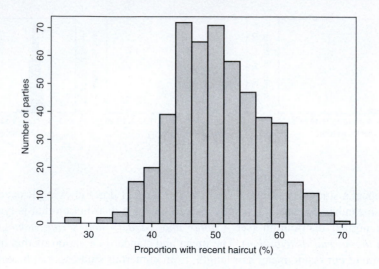

The histogram above actually couldn't be more fundamental to statistics. Remember that the data for each study don't follow the normal distribution. Instead of a smooth curve, you get a clump at 0 (didn't cut hair recently) and a clump at 1 (recent haircut), like this:

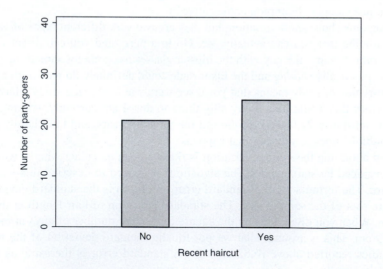

Yet a plot of the results of the study repeated multiple times gives a normal distribution (i.e., the figure at the top of the page). This is also true for the hair length example. The histogram of individual guys' hair length is skewed, with a few guys having hair much longer than average (I'll take my Fifth Amendment rights against self-incrimination.), but the histogram showing the mean hair length of repeated studies approximates to the normal. As a result, we can often use the normal distribution to calculate statistics even if the data are not themselves normally distributed.

• Things to Remember •

1. There are two sorts of variation
    a. One is natural variation that can be observed (such as hair length).
    b. The other is the variation of study results, when a study is repeated (which we don't often do).
2. Standard deviation is used to describe the natural variation of something you can measure.
3. Standard error is used to describe the variation of study results—a statistic such as a mean or proportion calculated from study data—imagining hypothetically you were to repeat a study many times.
4. The variation of study results often follows a normal distribution, even if the data from each individual study are non-normal.
5. If you drink at a party, don't drive.

> ⁘ **SEE ALSO**: *How to avoid a rainy wedding: Variation and confidence intervals*

for Discussion

1. What I am describing as a "study" here is when I go to a single party and measure the hair length of every guy. The aim of the study is to estimate the typical hair length of American men. What must I assume in order to use my data (hair length at one party) to inform my study aim (hair length of American men)?

2. What has statistics got to do with parties?

NOTE: See page 167 for answer sets.

CHAPTER 11

How to avoid a rainy wedding: Variation and confidence intervals

Rain in Santa Monica?

The weather in August in Santa Monica, California, hardly varies: it is always warm and sunny and rarely rains. August weather in London, where I grew up, is more variable and it often rains, sometimes for days at a time. This is why, when I got married in Santa Monica, I was panicking about putting up a tent to keep the rain off and my wife wasn't worried about anything. Confidence, it seems, is the opposite of variation: the more something varies, the less confident we are about what it is going to be.

Here is another example: my friend Jennifer is (almost) always in a good mood, so ask me, "How's Jennifer?" and I'll be pretty confident when I tell you that she's doing well. Another friend of mine (who shall remain nameless), is a lot of fun but moody—very moody. Ask me, "How's [name deleted]?" and the only reasonable answer is, "I haven't a clue." Again, more variation equals less confidence.

Making guesses about an individual

A game to play when your statistics lecturer hasn't shown up and you are out of things to do: try to guess whether the next student coming through the door is overweight. To keep things simple, we'll assume that we know that the student is male, but not an athlete (hence excluding the 300 lb offensive lineman). We'll also imagine that our task is to guess body mass index (BMI), which is weight in kilograms divided by the square of height in meters.

Now here is a bit of statistical information to help your guess. In a study of about 100 non-athlete male students at your university, the mean BMI was 26.0 and the standard deviation was 3.9. So if you had to guess the BMI of the guy just outside the door, your "best guess" would be 26.0. In fact, this is pretty much the definition of "average"—what you'd guess if you had to. But what if I asked you how confident you are that they have a BMI of *exactly* 26.0? Your answer should be "not very". Indeed, assuming that BMI is rounded to a single decimal place, it is possible to calculate that only about 1% of male non-athlete students have a BMI of exactly 26.0.

Your alternative to guessing 26.0 would be to say something like, "I guess that his BMI is somewhere between 22 and 30." How confident should you be in this answer? Well, 22 and 30 are about a standard deviation from the mean, and we know two-thirds of the data set are within a standard deviation of the mean, so you could say that you would be right roughly two-thirds of the time. If you guessed "between 18 and 34"—two standard deviations of the mean—about 95% of your guesses would be correct.

Making guesses about the results of a study

Let's try to do the same with the results of studies. The reason why your statistics teacher is late is because she is finalizing the analysis of BMI in 100 non-athlete males. What will she announce as the mean? Again, your best guess would be 26.0, but again you know that results of studies can vary. You also know that variation makes you less confident in your answer so, again, you'd rather give a range. But this time we aren't worried about how BMI varies between different individuals (the sickly-looking goth or the junk-food-addicted football fan) but how the mean BMI varies between studies (by chance, we end up sampling too many goths). So we want to think about standard error rather than standard deviation (see *Long hair: A standard error of the older male*). The standard error of our study is the standard deviation divided by the square root of the sample size, which gives 0.39. Our rule of thumb is that 95% of the results of a study (such as a mean) will be within two standard errors of the mean, so if we guessed that the results of the lecturer's study would be between 25.2 and 26.8 we'd have a pretty good chance of being right.

Confidence intervals and reference ranges

For reasons that are somewhat obscure, the range we give for values for an individual is called a *reference range;* the range we give for results of a study is called a *confidence interval.* A reference range is often used by doctors. For example, if levels of something you measure in the blood, such as white blood cells, are outside the reference range, the doctor concludes that the patient's white blood cell count is unusual and would consider some additional tests to see what might be wrong.

A confidence interval is useful for interpreting the results of a study. For example, imagine that we were looking at a study of whether a "mentoring" program affected Scholastic Aptitude Test (SAT) scores. We read that mentoring was associated with an increase in SAT scores by 4 points, with a 95% confidence interval of −2 to 10 (confidence intervals are almost always 95% confidence intervals). An obvious point is that it is plausible that mentoring actually makes things worse (an increase in SAT scores of −2 means a 2 point decrease in scores). So we certainly wouldn't recommend that schools start to implement mentoring programs. But we might also recommend that no more studies should be conducted on this particular mentoring approach. Our confidence interval tells us that it isn't likely that mentoring improves SAT scores by more than 10 points, which is a small level of benefit for a test scored out of 1600—certainly not enough benefit to start a whole new school program. On the other hand, if the confidence interval was to 54 we might conclude that, although we don't have clear evidence that the mentoring is effective, it might be, and further research should be considered.

If confidence intervals are so great, why should I (sometimes) ignore them?

A student of mine told me that if he had learned anything in my class, it was about the importance of confidence intervals. (As he put it, "You pounded it into me.") I felt a little bad about this because many of the confidence intervals I see in scientific papers are nonsense. As an example, I once read a study about men undergoing surgery for prostate cancer. The point of the study was to find out whether complication rates differed between the usual surgery, where the surgeon makes a big cut in the patient's abdomen ("open" surgery) and "keyhole surgery," which is when the surgeon conducts the operation using a special tube placed through a small hole in the patient's body (the "laparoscopic" approach). The authors reported that the mean age in the open group was 61 years, which is kind of useful to know. But they also reported a 95% confidence interval around this estimate of 60 to 62. What this tells us, roughly speaking, is the true mean age of patients undergoing open surgery for prostate cancer might not be exactly 61 but is unlikely to be much more than a year either side.

But I wasn't reading the study to find out the true mean age of the prostate cancer patients going for surgery—what I wanted to know concerned the relative complication rates. The authors reported that 2% of patients undergoing open surgery compared to 3% of those undergoing laparoscopy experienced a certain complication, a difference of 1%. They also reported a 95% confidence interval of −2% to 4%. What I can conclude is that the complication is rare and is unlikely to differ much between the different types of surgery (surgeons wouldn't consider 3 or 4% an important difference). For a different complication (which is fortunately less serious), rates were 30% in each group, a difference of 0%, with a 95% confidence interval of −9% to 9%. Now a surgeon would tell you that a difference in complication rates of close to 10% would be a big deal, so this means we can't rule out that open and laparoscopic surgery might have important differences in the rate of this complication.

On the other hand, if I was told that the chance of rain in Santa Monica in August was 2%, I'd be interested in the confidence interval around that rather than the confidence interval around the difference in weather between Santa Monica and London. This is because I'd already chosen to get married in Santa Monica, and I wanted to know the chance of having a rainy wedding, not the chance I'd made a poor decision.

This gives us a pretty easy rule of thumb: work out what it is you want to know from a study and then calculate a confidence interval around that. You might say that the indiscriminant use of confidence intervals in scientific papers is because the authors don't have a firm idea of what it is that they want to find out. And you might be right—I couldn't possibly comment.

• Things to Remember •

1. If we take a sample of individuals, 95% will have values within 2 standard deviations of the mean. This is called the reference range.

2. If we repeated a study a large number of times, 95% of the estimates—the mean BMI of students, or the difference in proportions between two types of surgery—would be within 2 standard errors of the true mean; this is called the confidence interval.

3. 95% of 95% confidence intervals will include the true value of an estimate. The true value of an estimate is called a parameter.

4. Reference ranges are only used for some very specific purposes, such as identifying patients with blood values suggestive of disease. As such, reference ranges are not usually reported in scientific papers.

5. Confidence intervals are a useful way of thinking what the results of a study might plausibly be, were you to repeat it.

6. It didn't rain at my wedding.

⁂ **SEE ALSO:** *Statistical ties, and why you shouldn't wear one: More on confidence intervals*

for Discussion

1. When we were trying to guess the body mass index of the student, I stated that 95% of the individual observations would be within about two standard deviations of the mean. I said something similar about reference ranges in the "Things to Remember." What is my assumption here?

2. When talking about the results of the lecturer's study on body mass index, I said that "95% of study results—the mean BMI—will be within two standard errors of the mean." What is the mean here? What is the standard error?

NOTE: See page 168 for answer sets.

CHAPTER 12

Statistical ties, and why you shouldn't wear one: More on confidence intervals

If I were to say the word "statistician" followed by the word "tie," you'd probably think of something brown and nylon. Or maybe gray. Either way, this is probably not the meaning intended when opinion pollsters tell you that two political candidates are "in a statistical tie." As it happens though, the concept of a "statistical tie" is an interesting way of looking at confidence intervals.

Having nothing better to do, one morning last fall I went out and asked 1000 people who they would vote for in the forthcoming election for the Senate: 483 respondents favored the Republican candidate and 516 favored the Democrat. (One guy said that he would be voting for a small, green alien from the planet Murg on the basis of intergalactic messages picked up through metal fillings in his teeth.) Let's say that I went for lunch and then polled another 1000 in the afternoon. Now we wouldn't expect to get precisely the same results. Indeed, if I found that exactly 483 of my afternoon sample were voting

Republican I'd probably be somewhat surprised. On the other hand, I'd probably also be surprised if results were very different. If, say, my afternoon revealed 10% support for each of the two major party candidates with 80% pulling for the alien, I'd conclude that I'd missed something on the news. (Or, alternatively, that I needed a new dentist.)

Most of us have a pretty good idea of what we would consider a surprising result for the afternoon poll. For example, 52% Democratic support seems reasonable and 40% does not. We can quantify what would be a "surprising" result by calculating a confidence interval.

As an illustration, I used a computer to simulate the results of my polls if the true level of Republican and Democratic support was tied at 50% each (we'll leave the Murgians aside for now). The following histogram shows the results of 100,000 simulations—in other words, the results I would get if I conducted my poll 100,000 times:

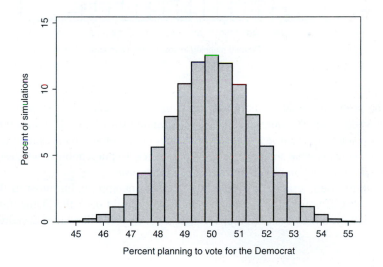

You can see from this graph that it wouldn't be that unusual to get the data I observed —51.6% pulling for the Democrat—if in fact the contest was 50:50. This is what pollsters mean by saying that the results of the survey are a "statistical tie": the results wouldn't be unusual if the true result were a tie.

Here is something else the opinion pollsters say, "We surveyed 1000 people and the results have a margin of error of plus or minus 3 percentage points." Now if you look at the graph, you can see that very few of my simulations had Democratic support at less than 47% or greater than 53%. In fact, 95% of the simulations have Democratic support between 47% and 53%. What the pollsters call a *margin of error* is similar to what statisticians call a confidence interval.

Back to those ties. Let's imagine that the true proportion of Democratic voters was indeed exactly 51.6%. We'd been absolutely correct that the Democrat was going to win, and had gotten the winning margin spot on. However, the pollsters had declared our results a tie. To think through what is going on, here is a histogram of the results of 100,000 simulations with the Democrat on 51.6%:

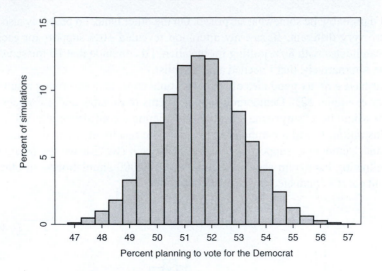

Part of the problem is that "a margin of error of 3%" is generally interpreted as "the true result is somewhere from 3% more to 3% less, we have no idea where." But as you can see from the graph, results close to the true result (51.6%) are more common than results that are 3% too high. The "true result" is what statisticians call a parameter. To put it another way, estimates close to the parameter are more common than estimates farther away.

Results with the Democrat on less than 50% are not uncommon, but most of the results do favor the Democrat. If you had to put money on it, you should back the Democrat. The confidence interval tells you that it isn't perhaps as sure a bet as you'd like but it certainly isn't a tie.

• Things to Remember •

1. You may hear the results of opinion polls described in terms of a margin of error. This is a concept similar to the confidence interval.

2. The idea of a margin of error makes it sound as though the true result could be absolutely anywhere within the confidence interval. In fact, it is more likely to be nearer the middle of the confidence interval than at either extreme.

3. A "statistical tie" means that the confidence interval includes no difference.

4. Put (2) and (3) together and you realize that a "statistical tie" doesn't mean, "I have no idea, might as well flip a coin."

5. I would like to state for the record that I do not own a brown tie (or a gray one).

SEE ALSO: *Avoid the sales: Statistics to help make decisions*

for
Discussion

• •

1. I argued that, if we had to bet on it, we should put money on the Democrat even though the confidence interval for the poll included the possibility that the Republican would win. Does this mean that we should just abandon confidence intervals then and go with whatever looks best?

2. Opinion pollsters for political races typically survey around 1000 people, but they don't go out in the morning and ask the first 1000 people they meet. What do they do?

_____

NOTE: See page 169 for answer sets.

CHAPTER 13

Choosing a route to cycle home: What *p*-values do for us

$$b = \frac{\sum (x - \bar{x})(y - \bar{y})}{\sum (x - \bar{x})^2}$$

SEE YA!

You have probably heard of the *p*-value. In fact, a common view is that the whole point of statistics is to produce *p*-values: those less than 0.05 are "statistically significant" and a good thing that makes everyone very happy, whereas "non-significant" *p*-values above 0.05 are bad and a source of shame. This begs the question of what *p*-values should actually be used for.

Going home each night I have a choice between cycling down a busy road or winding through the beautiful backstreets of brownstone Brooklyn. Being a statistician who thinks of nothing all day but inference and estimation, I have recorded how long each route takes me on a number of occasions and calculated means and standard deviations. Imagine that, one day, I had to get home as soon as possible for an appointment. To choose a route, I conduct a statistical analysis of my travel time data. It turns out that the busy road is quicker, but the difference between routes is not statistically significant ($p = 0.4$). Nonetheless, it would still seem sensible to take what is likely to be the quicker route home, even though I haven't shown convincingly that it will get me there fastest.

Now let's imagine that this incident got me fired up and I spend two years randomly selecting a route home and recording times. When I finally analyze the data, I find strong evidence that going home via the busy road is faster ($p = 0.001$), but not by much (it saves me 57 seconds on average). So I decide that, unless I am in a real rush, I'll wind along the backstreets simply because it is a more pleasant journey and well worth the extra minute.

We tend to think that p-values should determine our actions. For example, if a drug company conducts a clinical trial of a new drug, we tend to say, "$p < 0.05$ means use the drug; $p \geq 0.05$ means don't use the drug." Yet the bicycle example shows the opposite—I chose the busy road when p was 0.4 but not when p was 0.001. This suggests we need to think a little harder about what p-values tell us and how we should use them.

The most important thing to remember about p-values is that they are used to test hypotheses. This sounds obvious, but it is all too easily forgotten. Here are two examples (which are good ones, so I'll return to them). The first example is that you sometimes see scientific papers listing over 100 p-values. It generally isn't the case that the investigators really wanted to test that many hypotheses—they just got into the swing of things and kept asking their statistical software to spit out another p-value.

The second example goes back to clinical trials. In a trial of a new drug, whether a patient gets the drug or a placebo is determined at random—a bit like flipping a coin. This is called a *randomized trial*. The idea is that the patients getting the drug start off being similar to the patients on placebo, so any differences at the end of the trial can then be attributed to the drug. Flipping a coin to determine who gets the drug and who gets placebo makes it likely that you'll get similar numbers of, say, older people or women in each group. However, it doesn't provide any guarantees, so what a lot of researchers then do is run a check to see that everything balanced out by calculating a p-value for, say, the difference in average age between groups. This sounds sort of reasonable, but it actually makes no sense at all. The hypothesis being tested here is whether any apparent differences in age are simply due to chance or whether in fact one group is truly older than the other. But this is a randomized trial and we flipped a coin to determine who ended up in which group. So we *know* that any differences in age are due to chance. Anyone reporting a p-value for a baseline difference between groups in a randomized trial, and concluding that there were or were not "significant" differences for things like age or gender, has obviously forgotten that what we use p-values for is to test hypotheses.

It is also worth remembering that p-values are about testing hypotheses because, in many cases, this is not what we want to do at all. When I had to get home in a rush, I wasn't interested in knowing for sure which was the quickest way home, I just needed to work out what route was likely to get me to my appointment on time. Moreover, even when we do want to test hypotheses, our conclusion is a necessary but not a sufficient guide to action. I eventually provided strong evidence that using the busy road was quickest but decided to choose a different route on the basis of considerations—pleasure and quality of life—that formed no part of the hypothesis test.

• Things to Remember •

1. *P*-values test hypotheses.
2. That's it for this chapter.

:•: **SEE ALSO:** *The probability of a dry toothbrush: What is a p-value anyway?*

for Discussion

1. I stated that I "provided strong evidence that using the busy road was quickest." Why didn't I just say that I'd proved it?

2. We normally think that a big difference between groups means a small *p*-value. But I found a very small *p*-value ($p = 0.001$) even though the difference in travel times between the two different routes home was trivial. How come?

3. If statistics is not just about testing hypotheses, what else can you use statistics for?

NOTE: See page 171 for answer sets.

CHAPTER 14

The probability of a dry toothbrush: What is a *p*-value anyway?

I have a party trick: when I tell someone what I do, and they say, "statistician, eh? I took statistics in college," I ask them to define the *p*-value. (I know what you're thinking—not much of a trick. I'm working on some other stuff.) The point is, I have yet to meet anyone who has got anywhere close to the right answer. This is pretty odd because the *p*-value is such a key idea in statistics. Imagine if a literature graduate didn't know whether Shakespeare wrote plays or novels, or someone who'd taken an economics course couldn't describe the relationship between supply and demand. So, if you do nothing else, please try to remember the following sentence: "The *p*-value is the probability that the data would be at least as extreme as those observed, if the null hypothesis were true." Though I'd prefer that you also understood it—about which, teeth brushing.

I have three young children. In the evening, before we get to bedtime stories (bedtime stories being a nice way to end the day), we have to persuade them all to bathe, use the toilet, clean their teeth, change into pajamas, get their clothes ready for the next day and then actually get into bed (the persuading part being a nice way to go crazy). My five-year-old can often be found sitting on his bed, fully dressed, claiming to have clean teeth. The give-away is the bone dry toothbrush: he says that he has brushed his teeth, I tell him that he couldn't have.

My reasoning here goes like this: the toothbrush is dry; it is unlikely that the toothbrush would be dry if my son had cleaned his teeth; therefore he hasn't cleaned his teeth. Or using statistician-speak: here are the data (a dry toothbrush); here is a hypothesis (my son has cleaned his teeth); the data would be unusual if the hypothesis were true, therefore we should reject the hypothesis.

Statistical analysis of a set of data follows a very similar principle—you want to know the probability of the data if the null hypothesis were true. As an example, there was an idea a few years back that young people who had broken the law should be given tours around prisons to show them just how awful their lives would be if they didn't start behaving (that is, they would be "scared straight"). Some social science researchers decided to run an experiment in which young offenders were randomly chosen either to be scared straight or to be treated as usual in the criminal justice system (the control group). Here are some typical data from one of these experiments. 12 of 28 (43%) in the scared straight group committed a new crime compared to 5 of 30 (17%) in the control group.

It looks as though trying to scare teenagers makes things worse, though it could be that the study results were bad luck (in the same way that, by chance, you might beat me handily at Chutes and Ladders). To analyze the data statistically, we first write down a null hypothesis (which roughly speaking is that nothing interesting is going on). So our null hypothesis could be something like: "The chance of committing a crime after a first arrest is the same in teenagers going through scared straight as those going through the usual criminal justice procedures." Next, we conduct a statistical test and get $p = 0.043$. As the p-value is less than 5%, we call our result statistically significant, reject our null hypothesis and conclude that scared straight truly does make things worse.

Now, the p-value of 0.043 isn't quite the probability that, if there was no effect of scared straight, we would see exactly 12 of 28 committing new crimes in the scared straight group and exactly 5 of 30 in controls. This is because we also want to take into account the possibility that the results could have been reversed (that is, more crimes in controls) or might have shown an even bigger difference between groups (e.g., 100% crime rate after scared straight and 0% in controls). In both of these cases we would have rejected our null hypothesis.

Essentially what we do to get the p-value is to write down every possible result of the study (14 of 28 committing crimes in scared straight and 15 of 30 in controls, 1 of 28 crimes in scared straight, 29 of 30 in controls, 24 of 28 and 6 of 30, etc.). We then work out the probability of each result if the null hypothesis were true (e.g., a result like 1 of 28 crimes in scared straight vs. 29 of 30 in controls would be very unlikely if the arrest rate was truly the same in each group). Finally, to get the actual p-value, we add up the probabilities of all results that are at least as unlikely as the one we got.

We can also show this on a histogram. The x-axis gives the difference in arrest rates between the control and scared straight groups and the y-axis gives the probability of each possible result if the null hypothesis were true.

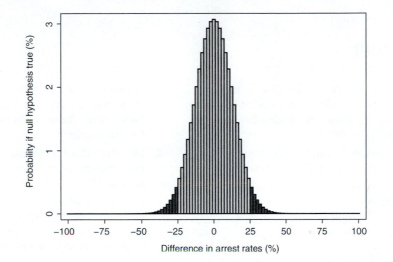

As you'd expect, the most common result if the null hypothesis were true is that there is no difference between groups, though small differences are also pretty common. Very large differences between the groups are extremely rare: the probability of a 100% difference—all controls being arrested and no scared straight arrests—is less than 1 in 10^{16} if the null hypothesis were true (not far from the chance of correctly identifying a randomly chosen grain of sand from all the beaches in the world). The difference we observed in the study was 26% and the shaded areas show results at least as extreme as this. If you add up all the shaded areas, what you get is the probability of getting a difference of 26% or more if the null hypothesis were true. This is the *p*-value: 0.043.

So here is what to parrot when we run into each other at a bar and I still haven't managed to work out any new party tricks: "The *p*-value is the probability that the data would be at least as extreme as those observed, if the null hypothesis were true." When I recover from shock, you can explain it to me in terms of a toothbrush ("The probability of the toothbrush being dry if you've just cleaned your teeth").

• Things to Remember •

1. Inference statistics involves testing a hypothesis, specifically, a null hypothesis.

2. A null hypothesis is a statement suggesting that nothing interesting is going on, for example, that there is no difference between the observed data and what was expected, or no difference between two groups.

3. The *p*-value is the probability that the data would be at least as extreme as those observed if the null hypothesis were true.

4. If the data would be unlikely if the null hypothesis were true, we conclude that the null hypothesis is not true.

5. My son has now worked out my trick and has taken to running his toothbrush under the tap for a second or two before heading to bed.

SEE ALSO: *Choosing a route to cycle home: What p-values do for us*

for Discussion

1. Why do we say the probability of the data under the null hypothesis? Wouldn't it be more interesting to know the probability of the hypothesis given the data?

2. Here are some research questions. Give an example of the null hypothesis for each of these:

   a. Does compulsory job retraining affect long-term unemployment?

   b. Do African American males have a harder time than white males hailing a taxi in New York City?

   c. Nationwide, about 28% of births are via Cesarean delivery. Do hospitals in New York State have higher than average Cesarean rates?

   d. Do after-school programs increase student participation in art, music, drama or dance activities?

   e. Do patients taking a new, less toxic type of chemotherapy have response rates at least as good as those on the standard (and unpleasant) chemotherapy drug?

NOTE: See page 172 for answer sets.

Michael Jordan won't accept the null hypothesis: How to interpret high *p*-values

So finally, after many hours of packing and loading, the bags are in the car, the children in their booster seats, the snack bag in easy reach, everyone is buckled in and my hand is on the ignition key. At which point my wife asks, "Where is the camera?" Being a statistician, I instantly convert this question into two hypotheses: "the camera is in the car" and "the camera is still in the house." Given that it is easier to pop back inside the house than to unload the car, I decide to test the second hypothesis. A few minutes later, I tell my wife that I have looked in all the normal places inside and couldn't find the camera. We conclude that "it must be in the car somewhere" and head off on our road-trip.

There is something a little odd behind this story: we concluded one thing (that the camera was in the car) because we couldn't find evidence to support something else (the camera was in the house). But as it happens, this is exactly what we do when we conduct a statistical test. First, we propose a null hypothesis, roughly speaking, that nothing interesting is going on (see *The probability of a dry toothbrush: What is a p-value anyway?*). We then run our statistical analyses to obtain a *p*-value. The *p*-value is the probability that the data would be at least as extreme as those observed, if the null hypothesis were true, so if the *p*-value is low (say, less than 0.05) we say, "These data would be unlikely if the null hypothesis were true, therefore it probably isn't." As a result, we declare our result "statistically significant," reject the null hypothesis and conclude that we do indeed have an interesting phenomenon on our hands.

As a simple example, we might have a null hypothesis that girls and boys learn handwriting at the same rate, and a data set of handwriting test scores divided by gender. A statistically significant *p*-value would lead us to reject this null hypothesis and conclude that there are differences in handwriting at an early age.

What we do if *p* is greater than 0.05 is a little more complicated. The other day I shot baskets with Michael Jordan. (Remember that I am a statistician and never make things up.) He shot 7 straight free throws, I hit 3 and missed 4 and then (being a statistician) rushed to the sideline, grabbed my laptop and calculated a *p*-value of 0.07 for the null hypothesis that I shoot baskets as well he does. Now, you wouldn't take this *p*-value to suggest that there is *no* difference between my basketball skills and those of Michael Jordan—you'd probably say something like our experiment hadn't *proved* a difference.

Yet a good number of otherwise smart people come to exactly the opposite conclusion when interpreting the results of statistical tests. Just before I started writing this book, a study was published reporting about a 10% lower rate of breast cancer in women who were advised to eat less fat. If this is indeed the true difference, low fat diets could reduce the incidence of breast cancer by tens of thousands of women each year—astonishing health benefit for something as simple and inexpensive as cutting down on fatty foods. The *p*-value for the difference in cancer rates was 0.07 and here is the key point: this was widely misinterpreted as indicating that low fat diets don't work. For example, the *New York Times* editorial page trumpeted that "low fat diets flub a test" and claimed that the study provided "strong evidence that the war against all fats was mostly in vain." However, failure to prove that a treatment is effective is not the same as proving it ineffective. This is what statisticians call "accepting the null hypothesis" and, unless you accept that a British-born statistician got game with Michael Jordan, it is something you'll want to avoid.

• Things to Remember •

1. Testing a hypothesis ("inference" statistics) consists of the following steps:
   a. Specify a null hypothesis.
   b. Apply a statistical test to the data to obtain a p-value.
   c. If the p-value is less than 0.05 ("statistically significant"), reject the null hypothesis.
   d. If the p-value is 0.05 or more, don't reject the null hypothesis.
2. Don't accept the null hypothesis. If the p-value is high, 0.05 or greater, you *can't* say, for example:
   a. There is no difference between girls' and boys' handwriting.
   b. Job training doesn't improve productivity.
   c. Andrew Vickers could have taken the Chicago Bulls to six NBA titles.

for Discussion

1. When discussing statistical significance, I have repeatedly described p-values less than 0.05 as "statistically significant" and p-values of 0.05 or more as "not statistically significant." Is it true that a p-value of 0.049999 is always statistically significant and that a p-value of 0.050001 is never statistically significant?

2. Why do you think that my experiment with Michael Jordan resulted in a non-significant p-value?

3. What should we conclude about the effects of a low fat diet on breast cancer?

4. What is the connection between a criminal trial and a p-value?

NOTE: See page 174 for answer sets.

The difference between sports and business: Thoughts on the *t* test and the Wilcoxon test

Sports is to business as Wilcoxon is to *t* test

I am sure you can track down some post-deconstructionist who'll tell you that sports is business by other means. (Or maybe the other way around.) But here's how I look at it: in sports, rank matters; in business, amount matters. In sports you want to come first; in business, you want to make a lot of money. The Mets might have a mediocre season, ending only a few games over .500, but if they are first in their division they'll go to the playoffs—end of story. The owners of the McAllister Towing and Transportation Company don't really care whether they are the most profitable tugboat operation in the New York area—they just want to make a profit and it will hopefully be more than a few dollars past breaking even.

As regards statistics, the Wilcoxon is like sports and the *t* test is like business. Here is how a Wilcoxon test works: you compare whether ranks are higher in one group than the other. Here is how a *t* test works: you compare whether the mean is higher in one group than the other group.

Analysis of a sports experiment

In the Tour de France, each team is accompanied by a *soigneur*, or "healer," whose job is to give massages to help the cyclists recover from each day's ride. Imagine that you are the coach of the Columbia University cycling team and that your team has picked up a volunteer *soigneur* from a local massage school. This seemed like a good idea at first, but you are now rather tired of the *soigneur* interrupting training to discuss mystical energy flow and you want to find out for sure whether massage actually helps. So you run the following experiment: on Sunday, your team takes part in a 50 mile race. That afternoon, you randomly select half of your team to get a massage. On Monday, you send everyone for a time trial. You then look at the time trial data to address the null hypothesis that the cyclists receiving the massage were no quicker than those who did not receive a massage.

Here are the results, with the times given as minutes:seconds:

Massage group	Control group
51:55.1	48:49.9
53:39.7	53:17.4
58:29.8	59:33.6
59:22.8	60:49.4
59:24.1	61:12.7
59:57.2	62:33.6
60:32.1	63:18.7
61:43.3	63:19.2
63:13.4	65:15.5
63:40.3	65:25.4

To do a *t* test with these data, you work out the mean and standard deviation in each group. You get a mean of 59.20 (SD 3.788) minutes in the massage group and 60.36 (SD 5.337) minutes in the control group. If you plug these numbers into the appropriate formula, you get a difference between groups of 1.16 minutes. This means that, on average, cyclists receiving a massage completed the time trial about 1 minute and 10 seconds faster than those in the control group. The standard error of this difference is 2.07. We know that, if the null hypothesis were true, then 95% of the time the difference between groups would be no more than two standard errors away from zero. For this experiment, we are about half a standard error away from zero and so we know we have a non-significant result (it is actually $p = 0.6$) and can't reject the null hypothesis of no difference between groups.

To do a Wilcoxon, you have to work out the ranks of the data—who came first, who came second and so on. The cyclist coming in first was in the control group with a time near 49 minutes, and this individual gets a rank of 1. The next fastest, a full three minutes back, was a cyclist in the massage group, who gets a rank of 2. If you keep going assigning ranks in terms of where each cyclist came in the time trial, what you get is something like this, with the ranks shown in the brackets:

Massage group	Control group
51:55.1 (2)	48:49.9 (1)
53:39.7 (4)	53:17.4 (3)
58:29.8 (5)	59:33.6 (8)
59:22.8 (6)	60:49.4 (11)
59:24.1 (7)	61:12.7 (12)
59:57.2 (9)	62:33.6 (14)
60:32.1 (10)	63:18.7 (16)
61:43.3 (13)	63:19.2 (17)
63:13.4 (15)	65:15.5 (19)
63:40.3 (18)	65:25.4 (20)

If you add up the ranks in each group, you get a total of 89 in the massage group and 121 in controls, a mean of 8.9 and 12.1. So, the average cyclist receiving a massage would come in 9th—three places ahead of the average cyclist not receiving massage, who'd come in 12th. (Ok, I know—you can't get a whole bunch of cyclists all coming 9th in a time trial, but you take my point.) Our question now is whether the difference in average rank of 8.9 and 12.1 is statistically significant. To get a p-value, we can think back to the definition of the p-value in terms of the probability of the observed data or something more extreme—if the null hypothesis were true. What we'd expect if the null hypothesis were true is that there would be no difference in rank between the two groups. But we wouldn't be surprised if there was sometimes a small difference in ranks and occasionally we'd expect see a big difference, just by chance. This figure shows the difference in ranks if the null hypothesis were true and massage didn't have any effect on cycling performance:

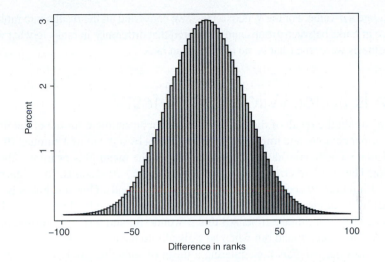

As you can see, you get a normal distribution. I've shaded all bars where the difference in ranks was as high or higher than what we saw in our study, which was 32. You can see that it isn't particularly unusual to get a difference in ranks that large and so we'd guess that we don't have a statistically significant difference. If you add up the height of the shaded bars (the *p*-value is the probability of the observed data or *something more extreme*, meaning a bigger difference in ranks), you get 0.113. But remember, we have to take into account the possibility that we might get better results in the control group, and that this would also lead us to reject the null hypothesis. When you add up the height of the bars for a difference in ranks of −32 or less, you get, naturally, 0.113. Adding 0.113 and 0.113 gives a total probability of 0.226. So we can state that there is a 22.6% chance that you would see a difference in ranks of 32 or more if the null hypothesis were true, that is, no effect of massage on cycling times. This is the *p*-value you get when you run a Wilcoxon test on these data.

So you should fire the soigneur, right?

The answer to that question probably depends on just how annoying you found all that talk of energy chakras. The key thing from a statistical point of view is that our non-significant *p*-values don't mean that we accept the null hypothesis. The fact that the data would be reasonably likely if the null hypothesis was true doesn't mean that the null hypothesis is true, after all, the data would also be reasonably likely under a hypothesis that massage did improve cycling times, but only by 15 seconds.

The key thing to remember for now is that the cycling experiment helps explain the difference between the *t* test and the Wilcoxon. For the *t* test, we first calculated an estimate that addressed our research question: we wanted to know if massage improves cycling times and so we calculated that, on average, cyclists receiving the massage were quicker than controls by a little over a minute. We then calculated a standard error for this estimate and divided one by the other to calculate a *p*-value. For the Wilcoxon test, we converted all the results into ranks, calculated the difference in ranks between groups and compared that difference in ranks to what we'd expect if the null hypothesis were true (that is, no difference in rank).

So what is better, Wilcoxon or *t* test?

The best thing about the result of the *t* test is that we got an estimate for the effect of massage on cycling performance: massage improved time trial times by a mean of 1 minute 10 seconds. We can also calculate a 95% confidence interval, which is the mean plus or minus about twice the standard error. The standard error was around 2 minutes and 10 seconds, so we get a confidence interval of −3 minutes 10 seconds to 5 minutes and 30 seconds. Our best guess is that massage reduces times by a minute or so, but it could actually slow you up by 3 minutes or lead to a dramatic 5 minute reduction in race time. All this is wonderful and terrific information, but only if it is correct. As Bill Gates found out when he walked into the diner (see *So Bill Gates walks into a diner: On means and medians*), calculating a mean of something only makes sense in certain circumstances. You'd hardly want a *t* test to compare the difference in salaries between Dizzy's diner (where Bill is tucking into a three egg omelet) and the Little Purity diner down the block (which has no visiting billionaires) because you'd calculate a difference in mean salary of $125m, with a 95% confidence interval from −$125m to +$375m. As you can tell from this example, the *p*-value you get from a *t* test is not very reliable when the data are very skewed—in simple terms, the *p*-value is too high if there is a true difference between groups.

Moreover, although I suggested that getting an estimate is a good thing (it normally is), estimates aren't always interesting. If a biologist has some complicated hypothesis about the effects of a gene, and conducts a mouse experiment, the estimate might be something like "IL-2 production from PMA stimulated CD45RA + CD4 + cells was increased by 0.002 units in knockout mice." It is unclear whether this has any meaningful interpretation. The main point of such laboratory experiments is to investigate hypotheses and, as such, it is only the *p*-value that is of interest.

So, again, it all depends. This makes sense, because if you could really say whether the *t* test or Wilcoxon was better, whichever one was worse wouldn't be used anymore and I wouldn't have to write chapters about it.

• Things to Remember •

1. Statistical tests are applied to data to generate *p*-values to test hypotheses.

2. The *t* test and the Wilcoxon test are two well known statistical tests.

3. Both tests are used when two groups (such as boys and girls or treatment and control) are compared with respect to a continuous variable (such as time to complete a cycle race).

4. The *t* test involves calculating an estimate addressing the hypothesis of the study as well as its standard error. The *p*-value is calculated by comparing the estimate to the standard error.

5. To run a Wilcoxon test, the data must first be converted to ranks. The *p*-value is calculated by comparing differences in ranks to an expected distribution of differences in ranks.

6. The *t* test can be unreliable if the data are very skewed.

for
Discussion

1. You may have heard that statistical tests come in one of two flavors: parametric and non-parametric. The *t* test is a parametric test; the Wilcoxon is non-parametric. What does "parametric" mean and why is the *t* test, but not Wilcoxon, parametric?

2. What would you conclude from our experiment about the value of massage for recovery from cycling?

NOTE: See page 175 for answer sets.

CHAPTER 17

Meeting up with friends: On sample size, precision and statistical power

Why meeting up with friends can get so complicated

Say you want to hang out with your friend Doug, and he says, "Movie?" and you say, "Drinks?" and he says, "The Loki Lounge?" and you say, "9 pm?" and then there you are hanging out a couple of hours later. But then next time he suggests dinner with his friends Launa and Tom, and you want to go to Maria's bistro, but Tom doesn't like Mexican and wants to go to Café Steinhof instead, but you don't like *schnitzel*, at which point you remember a movie you really want to see, but Doug's seen it already, so Tom suggests a different movie, which Launa doesn't fancy and anyway, she wants dinner, and 20 minutes later the four of you are still discussing where to go and thinking about meeting up for coffee. As you probably know, the hassle of organizing something rises exponentially with the number of people you have to organize, following the famous equation $E = mc^2$ where E is effort, m is the mean fussiness (or flakiness) of your friends and c is the size of the crowd.

The $E = mc^2$ of statistics

Statistics works in a similar way, but sort of in reverse: each person you add to a group of friends planning to hang out results in exponentially more complications; adding another person to a group of study participants results in exponentially less information. The equivalent of $E = mc^2$ for statistics is:

$$\text{Precision} = \frac{\text{Variation}}{\sqrt{\text{Sample size}}}$$

Precision here can be thought of the width of your confidence interval. If you conduct a study and have a wide confidence interval, then your results are not precise. For example, if you obtained voting intentions from 20 likely voters, you might report that "55% of voters favor the Republican, with a 95% confidence interval of 32% to 77%" which is to say, you really have no idea whatsoever who is going to win.

In the formula, *variation* refers to the natural variation of whatever it is you are studying. In a study looking at a continuous variable such as age, height or blood pressure, that variation is expressed in terms of standard deviation: a large standard deviation of say, age, means that the individuals in your study are all sorts of ages from young to old (e.g., a study of voters); a small standard deviation means that most of those in your study are a similar age (e.g., a study of professional football players). In a study of a proportions, the variation depends on how close the proportion is to 50%. Think of it this way: if you put 1000 American voters in a room and asked who voted for the current president—this is typically somewhere close to 50%—you would find a lot of variation—many did but many didn't. If you asked how many had gone skydiving, there wouldn't be much variation because most people avoid throwing themselves out of airplanes.

The key point is that you can't do much about variation—it is just out there as a natural part of the world. So if you want to make your results more precise, you need to increase your sample size: put more patients on a clinical trial, interview more voters about their voting intentions, include more schools in an education study, and so on. But precision is related to the square root of sample size, so if you want to double your precision you need to quadruple your sample size.

As a simple example, imagine that you surveyed 100 students about their views on a recent change in how exams were organized and 38% were in favor. The confidence interval is about $\pm 10\%$ (i.e., 28% to 48%). If you conducted a survey with four times as many students and obtained similar results, your confidence interval would be half as wide ($\pm 5\%$ or 33% to 43%).

The inverse square law and hypothesis testing

If you are in the business of estimation, the *inverse square law* gives you a nice rule of thumb. Things are a little more complicated for inference. Imagine that you wanted to test a null hypothesis that "50% of students are in favor of the change." If you rejected this null hypothesis, it might allow you to say, for example, that most students were against the change and lobby the administration to change things back to the way they were. To think through the effect of sample size on hypothesis testing, I'll do what I have done before (see *Long hair: A standard error of the older male*) and use a computer to simulate the results of a very large number of studies, each conducted with 100 students and assuming that, in truth, only 38% were in favor of the change to the exams. In previous chapters, I have shown the distribution for the mean or proportion in order to explain ideas such as the standard error. What I show here is the upper bound of a 95% confidence interval:

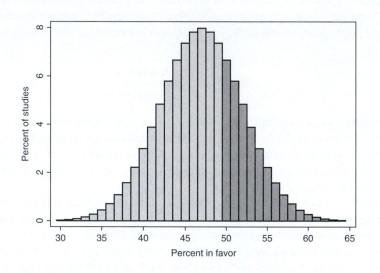

One thing we know is that if the upper bound of a 95% confidence interval excludes 50%, we can reject the null hypothesis that "50% of students are in favor of the change." This is because, if the upper bound is less than 50%, then the lower bound must also be less than 50% and, generally speaking, a confidence interval that does not include the null indicates a statistically significant result. What you can see from the graph is that the upper bound would be higher than 50% about one-third of the time (the shaded area). The null hypothesis is false—the proportion of students in favor is 38%, not 50%—and yet we fail to reject the null hypothesis pretty often. In this respect, a statistical test is similar to a diagnostic test (see *I ignore my child's cough, my wife panics: About specificity and sensitivity*) in that it doesn't always give you the correct answer. Just as you might be told that you're sick when you're fine or get a clean bill of health when you actually have some kind of disease, a test might give $p \geq 0.05$ when the null hypothesis is false and $p < 0.05$ when it is true.

Take a test that a doctor might give you for strep throat:

		True diagnosis	
		Strep throat	**Not strep throat**
Result of doctor's test	Positive	True positive	False positive
	Negative	False negative	True negative

Now imagine that instead of a diagnostic test, we had a statistical test. We'll call rejecting the null hypothesis a "positive" result. ($p < 0.05$ is normally seen as a good thing, right?)

		Truth	
		Null hypothesis false	**Null hypothesis true**
Result of statistical test	Statistically significant: reject null hypothesis	True positive	False positive
	Not statistically significant: fail to reject null hypothesis	False negative	True negative

The table shows that when you run a statistical test, you can only make one of two errors: you can reject the null hypothesis when it is true (a false positive) or fail to reject the null hypothesis when it is false (a false negative). Statisticians call this type I and type II error, respectively (helpful, huh?). Another name is α (alpha) error for false positive and β (beta) error for false negative. I introduced α in the discussion section for *Michael Jordan won't accept the null hypothesis: How to interpret high p-values*. Although we normally describe a *p*-value < 0.05 as "statistically significant," you can choose other values against which to compare your *p*-value. For example, you can specify before you analyze your data that *p*-values less than 0.01 will be deemed statistically significant and lead you to reject the null hypothesis. This is the same as saying "our level of alpha is 0.01."

Alpha is false positive, rejecting the null hypothesis when it is in fact true. Beta is false negative, failing to reject the null hypothesis even though it is false. Back to our student survey: we worked out that, about a third of the time, the upper bound of a 95% confidence interval would be greater than 50%, and thus we would inappropriately fail to reject the null hypothesis. In other words, beta, our false negative rate, for this study is around 0.3. As it happens, statisticians tend to talk about "power" rather than beta: power is 1 – beta, or simply the true positive rate.

A big difference between power and alpha is that alpha is something that you just decide on (it is almost always 5%). Power, like precision, depends on variation and sample size. Variation is a natural part of the world and you can't change it, but you can choose a sample size for your study. A statistician would describe our student survey in the following terms: "We will use an alpha of 5% and test the null hypothesis that 50% of the students are in favor of the new exams. Assuming that the true proportion is in fact 38%, a survey with 100 respondents will have a power of around 70%." In other words, if all your assumptions were correct (the true proportion is 38%; you manage to get 100 answers for your survey), you have a 70% chance of rejecting the null hypothesis at a significance level of 5%.

We can then use the computer simulation to imagine that we repeated a similar study but with 400 students. What we get is:

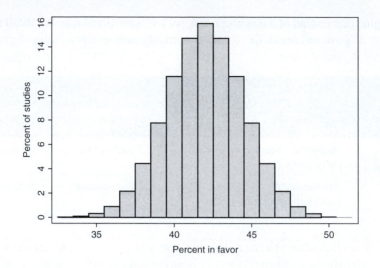

In this case, you very rarely see an upper bound for the confidence interval above 50% and therefore almost always reject the null hypothesis. The power of a study with 400 respondents is in fact greater than 99%. You can see that, to work out the effects of a change in sample size on hypothesis testing, you need to take into account the shape of the normal distribution. Indeed, the formula for power includes numbers that you have to look up using the normal distribution.

Calculating sample size

We can't use our simple inverse square law to work out power. But, as it turns out, we don't typically try to work out power at all. The typical question that scientists ask isn't, "OK, I've got 100 people here. What are my chances of a positive result?" but, "I want to test a hypothesis. How many people will I need?" This sort of question is particularly important in medical research. In a trial of a new drug, for example, you want to have a good chance of a statistically significant result if the drug is effective because it would be great to have another way to help sick people. But you can't have too large a sample size—drugs often have side effects, so you don't want to give a new drug to lots of people if it doesn't work. A typical question asked of statisticians in medical research might be something like: "About 50% of patients recover from a typical cold within 48 hours. We think that our new cold treatment might increase this to 70%. We will use the usual alpha of 5% and a power of 90%, that is, if the drug is effective, we want a 90% chance of showing that it does indeed work. How many patients do we need?"

To answer this question the statistician uses a formula that includes numbers taken from the normal distribution, but in its general form it is very simple. You might have heard of the "signal to noise" ratio: the more noise you have relative to the signal, the harder it is to hear. It is exactly the same for medical research, except that "harder to hear" means you need a large sample size and, as you might have guessed, the inverse square law applies.

$$\text{Sample size} = \left(\frac{\text{Noise}}{\text{Signal}} \right)^2$$

This is pretty similar to the first formula, except that we have *signal* instead of *precision* and *noise* instead of *variation*. Noise is in fact the same concept of variation and signal refers to the size of the difference from the null, often called the effect size. In the drug trial example, the drug was thought to increase recovery rates from 50% to 70% and so the effect size was 20%. This gets us back to the inverse square law. If you do the right calculations, it turns out that the number of patients needed for the drug trial is a little over 250. Now let's imagine that we wanted to see whether the drug might improve 48 hour recovery rates from 50% to 60%. Our effect size is now 10% and, according to our formula, if you halve the signal you quadruple the sample size. If you run the numbers for a trial with an alpha of 5%, an expected rate of 50% in the control group and 60% in the experimental group, a total of a little over 1000 patients are needed. The inverse square law isn't always exact—which is why you should use the proper statistical formula to work out sample size exactly—but it is a good rule of thumb.

The difference between theory and practice

In theory sample size, precision, effect size and statistical power are really quite straightforward. To work out power or precision you use a formula; the relationship between precision and sample size, and also that between effect size and sample size, follows the inverse square law. That is, in theory. But you know the difference between theory and practice, right? In *theory*, there isn't one.

The reason why these sorts of calculations are so difficult in practice illustrates a very important principle of statistics: statistics involves a bunch of formulas, but what you get out of those formulas depends on what you put in.

Here is an example typical of my day-to-day work. Some doctors ask me for help designing a study of a new drug for pain after surgery. They tell me that with the usual drug, pain is usually around 5 on a 0–10 scale and that they'd like to see this go down to 4 with the new painkiller. They also tell me that the standard deviation is about 1. When I plug these numbers into a sample size formula (using the typical alpha of 5% and a power of 90%), I get a total sample size of 44. At this point everyone gets excited because this isn't a large number of patients, so we can get the trial done by Christmas and have the paper published in some important medical journal well in time for the departmental review next year. But then a colleague points out that the drug is very safe and inexpensive and would be worth giving if it reduced pain scores by only half a point; moreover, didn't that recent scientific paper on postoperative pain report a standard deviation of 2? Now when I run the numbers I get a sample size of 674. A trial that large just isn't feasible at our hospital, which leaves the doctors rather depressed about their promotion prospects. However, one of the doctors is still hopeful and gets up enough energy to kick off what one statistician has called "the sample size samba": we can't possibly do a trial with 674 patients, but hang on, here is a paper saying that standard deviation *is* actually 1, so why don't we split the difference and call it 1.5? Oh, we'd need 380 patients, which is still too many. What if we change the difference between groups to 0.75? And now the sample size calculation spits out 170 patients, which is just about doable (if not in time for departmental review), so we agree on that.

This might make you rather skeptical of sample size calculation. But before you throw it away altogether, here is a little story. A surgeon specialized in a procedure that was usually pretty successful, with only 4 or 5% of patients experiencing a recurrence of their disease. Nonetheless, the surgeon thought that a slight variation in the way that the patient's lymph nodes were treated might affect outcome. The surgeon worked at a big hospital and wanted to go back and examine the case notes from all 1200 patients ever operated on at the hospital to see if this was indeed the case. Now if the best method of node removal led to a 10% relative decrease in recurrence rates, that would obviously be important to know. However, detecting a difference of this magnitude would require close to 60,000 patients at least, far more than the surgeon had available.

In short, formal sample size calculation can help us think through what research can and cannot tell us. There are lots of things we'd like to know, but if we only have a small sample, or are looking for very small effects, it is unlikely we'll ever find out for sure. Thinking about what we can and can't find out—and what we should do in the absence of clear evidence—could not be more central to the scientific process.

• Things to Remember •

1. When you are planning a study, you have to work out the sample size you'll need.

2. If the main purpose of your study is estimation, then the sample size you'll need depends on how precise you'd like your estimate to be. Precision can be thought of in terms of the width of your confidence interval.

3. Precision follows the inverse square law: if you want to halve the width of your confidence interval, you have to quadruple your sample size.

4. If the main purpose of your study is hypothesis testing, then the sample size you'll need depends on power. Power is the probability that, if there is truly an effect of a given size, you will reject the null hypothesis.

5. Power is typically set at 80% or 90%.

6. There are formulas to work out sample size exactly.

7. You have to be very careful about the numbers you put into sample size formulas if you want the numbers that come out of them to be sensible.

---

for Discussion

1. When calculating the sample size needed for a study with a hypothesis test, you need to specify an effect size, the difference from the null that you want to find. How do you choose an effect size?

2. Something that often happens is that a study fails to reject the null hypothesis. This typically leads the investigators to start running around to find someone to blame for this "negative" result. A common question is: what was the power of the study anyway? Is this a sensible question to ask?

3. Should you always do a sample size calculation when planning a study?

4. *For enthusiastic students only:* When we looked at the power of our survey study, I showed a distribution for the upper bound of the 95% confidence interval. I said that when this bound was less than 50% (our value for the null hypothesis), the result would be statistically significant. As such, I claimed that this distribution was the same as that for statistically significant results. Is this actually true?

NOTE: See page 177 for answer sets.

CHAPTER 18

When to visit Chicago: About linear and logistic regression

I was once asked to justify why I thought that a particular cancer study needed exactly 18 patients. I got no further than saying that, in simple terms, 18 was near 16 and the square root of 16 was 4 before I was interrupted. One of the world's leading cancer biologists told me that my explanation was getting too complicated and that, never mind, he'd just trust me. Which made me think that the most common type of regression I have to deal with at work is not linear or logistic regression—important statistical techniques both—but *infantile* regression.

Why linear regression leads to infantile regression

Here is a table giving the average temperature in Chicago:

Month	Average temperature in °F
January	21
February	25
March	37
April	49
May	59
June	69
July	73
August	72
September	64
October	53
November	40
December	27

You can see that, roughly speaking, the temperature increases by about $10°$ a month from January to June and then falls by about $10°$ a month from August to December. This means that, to get a pretty good estimate of the average temperature in Chicago for any particular month, just count how many months you are from summer (June–August), multiply by 10 and take that number from 70. October is 2 months way from the end of the summer (August), so that gives $70 - (2 \times 10) = 50°$; January is 5 months away from the beginning of summer (June), so that gives $70 - (5 \times 10) = 20°$. This isn't exact—the actual mean temperature is $53°$ in October and $21°$ in January—but it gets you pretty close and provides a good rule of thumb (which is to avoid Chicago in January).

There, that wasn't so bad was it? What we just did was a regression. Regression, it seems, has a particular ability to reduce otherwise emotionally healthy adults to an infantile state, blubbing hysterically and looking for someone's hand to hold. My guess is that this suits most statisticians just fine—a textbook on regression might look like a bunch of formulas to you; to statisticians like me, it is 450 pages of job security.

Here are some things that you might read about in that regression textbook: "the model sum of squares divided by the mean square error follows an F distribution"; a good model "increases the log likelihood"; "residuals" can be used to identify overly influential observations; it can be important to check for "heteroscedasticity." Here is what you need to know about regression: it is just about x and y. As in $y =$ average temperature; x is months away from summer; $y = 70 - 10x$.

Your math teacher was right (again)

What you might remember from high school math (or indeed *Why your high school math teacher was right: How to draw a graph*) is that you draw a graph by writing out an equation such as $y = 3x + 4$. You could also have something a little more complicated like $y = 2x^2 - 3x + 4$. One way of thinking about this is "give me a value for x and I'll work out y for you." In other words:

- y is something we don't know and want to predict.
- x is the information we have.
- the equation tells us how to work out y from x.

What regression does is work out what the equation should be. Keep saying to yourself "regression gives you an equation" and it will all seem a lot more manageable.

How fast can I run 26 miles?

I am hoping to run in the New York Marathon for the first time next year and it would be interesting to know how long it is likely to take me. Now if you had to guess, you'd probably start at the mean marathon time (one definition of the mean is what you'd guess if you had to—see *How to avoid a rainy wedding: Variation and confidence intervals*). I have a data set of marathon times and the mean running time is 4 hours and 5 minutes, or 245 minutes. Remember that y is what we want to know, the running time. So, we now have a regression equation:

$$y = 245$$

This is a start, but a pretty basic one. The problem is that we are not taking into account any other information. My guess is that a young man who runs 50 miles a week would run a faster time than an older woman who trains less intensively, but our equation would give both of them exactly the same time: 245 minutes. When I look at the data set, I find that the mean time for women is 263 minutes and the mean time for men is 239 minutes, a difference of about 24 minutes. Now that I have an x, some information to help work out y, I can now update our regression equation. If I use $x = 1$ for women and 0 for men I get:

$$y = 24x + 239$$

Eric and Erica have a shouting match

Delighted with this statistical analysis, I leave work and head to a party where I meet a woman named Erica. She tells me that she is running the marathon and wants me to predict her time. I say, "Woman, so x is 1, and so I guess your running time at $239 + 24 \times 1 = 263$ minutes." Our conversation is overheard by a guy named Eric, who is also planning to run the marathon and asks the same question. I give him a predicted time of $239 + 24 \times 0 = 239$ minutes. At this point, Erica becomes annoyed. "*Him*? Run faster than *me*? I train 40 miles a week, he runs a couple of five milers!" This seems a fair point: perhaps our regression equation would be better off if we tried to work out running time based on weekly training miles rather than gender.

Unlike gender, which takes two values, there are a lot of different possible values for weekly training miles, so we can't work out x just by taking means. What we do is plot weekly training miles against race time:

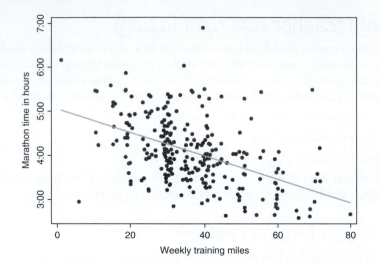

Each dot is an individual runner. You can see, for example, that there is one runner who trains 80 miles per week and ran the marathon in slightly under 3 hours. There is also someone who appears to be able to run a 3 hour marathon without ever really training; this might be someone who misunderstood the question—perhaps he thought we were asking about length of a typical training run. Either way, it is pretty typical to see a couple of data points that are difficult to explain. The graph also shows a line that fits through the dots on the graph. The equation of the line is:

$$y = -1.59x + 303$$

As you might guess, this is a regression equation for training miles and race time. Indeed, the line on the graph is often referred to as a *regression line*. But it is just high school math, right?

Multivariable regression

The problem now is that we'd predict Erica would run faster than Eric, and this sends Eric off on a riff about being a foot taller than Erica and at least being able to run a full stride rather than waddling along like an oversized duck, and there being no way that as a 25-year-old he would wouldn't be able to keep up with someone approaching retirement age. Eric has a point, which is that we are better off with more information, rather than trying to boil everything down to training miles. If Eric was a statistician, he'd say that in place of *univariate* regression, where we use one x to predict y, it is generally preferable to try *multivariable* regression, where we use several x's (x_1 for gender, x_2 for age, x_3 for training miles). If we run a multivariable regression on our marathon time data set we get:

$$y = 23x_1 + 0.81x_2 - 1.57x_3 + 262$$

To explain a bit of jargon, the numbers before the x's are called *coefficients* (statisticians sometimes also use the term *betas* or β's). So you might hear it said that "the coefficient for age was 0.81 per year." This means that our prediction for race time increases by 0.81 minutes for each one year increase in a person's age. The coefficients can also be thought of as the *slope* of the regression line. This is easiest to see in the graph of race times against training miles: the regression line has a slope of -1.59 because the equation for the regression is $y = -1.59x + 303$.

Using the formula, we can calculate the following predictions for Eric and Erica.

Eric: is male ($x_1 = 0$), 25 years old ($x_2 = 25$) and trains 10 miles a week ($x_3 = 10$). This gives $(23 \times 0) + (0.81 \times 25) - (1.57 \times 10) + 262 = 267$ minutes, or 4 hours 27 minutes.

Erica: is female ($x_1 = 1$), 48 years old ($x_2 = 48$; Eric was exaggerating) and trains 40 miles a week ($x_3 = 40$). This gives $(23 \times 1) + (0.81 \times 48) - (1.57 \times 40) + 262 = 261$ minutes, or 4 hours 21 minutes.

The two times are pretty close, but we predict that Eric's youth is not going to make up for his lack of training and reckon that Erica will beat him. On the day of the race, Erica comes in at 4 hours 19 minutes (259 minutes). Eric won't tell me his time for a while, explaining that he messed up his pacing and got a blister, but finally admits a slow 4 hours and 40 minutes (280 minutes). Our predictions weren't that precise (we certainly weren't able to predict that Eric would get a blister), but here is the key point: the predictions from the regression equation were closer to Eric and Erica's true race times than was the mean race time for all runners. If we had used the mean race time (245 minutes) we would have been wrong by 14 minutes and 35 minutes for Erica and Eric respectively. Instead, we were wrong by 2 and 13 minutes. Moreover, we were able to predict who came in first.

Logistic regression

Marathon running is a nice example of regression but, as you might guess, statisticians don't spend much time trying to guess running times. Something I do in my everyday work is calculate regression equations to predict whether a man has prostate cancer. Here is an example:

$$y = 0.121x_1 - 13.6x_2 + 0.020x_3 - 0.87$$

Here, x_1 is the blood level of a protein, prostate specific antigen (PSA); x_2 is the ratio between the total amount of PSA and the amount of PSA that is not bound to a second protein in the blood; x_3 is the patient's age in years. In other words:

$$y = 0.121 \times \text{PSA level} - 13.6 \times \text{Ratio of free-to-total PSA} + 0.020 \times \text{Age} - 0.87$$

Using our equation, if Mr. Smith was a 58 year old man with a PSA of 4 and a free-to-total PSA ratio of 25%, we would calculate y to be $(0.121 \times 4) - (13.6 \times 0.25) + (0.020 \times 58) - 0.87 = -2.63$. It isn't clear what exactly this means. For the marathon running example, y was the time in minutes—a continuous variable—so we call the regression a *linear* regression.

In the case of cancer, however, you either have it or you don't. Regression for binary variables such as cancer is called *logistic* regression.

Now statisticians don't go around saying, "According to my equations, Mr. Jones has cancer and Mr. Smith doesn't." What we do is give the *probability* that Jones or Smith has cancer. As it turns out, y isn't quite a probability, it is the log odds. I won't explain the log odds for right now other than to say that it can be easily converted into a probability. Mr. Smith had a y or log odds of -2.63, which converts to a probability of about 7%. This is pretty low, so his doctor would probably reassure him and tell him he had no need of further tests.

We can play with these numbers a bit. If Mr. Smith was a bit older, say, 65, this doesn't change his risk much (it goes up to 8%); if PSA was higher (say, 10), risk does increase (to 13%), but not enormously; however, if the ratio of free-to-total PSA was lower (say, 15%), this has a big impact on cancer risk (risk is now 22%).

This tells us something very important: I introduced regression in terms of prediction (can we predict marathon time? Can we predict prostate cancer?). But if you can predict something, you often understand it pretty well. What matters for marathon running is less whether you are a man or a woman, or how old you are, but how many miles you run in training; what matters for prostate cancer is not so much how much PSA you have in the blood, but whether that PSA was "free" or bound to another protein. This is one of the reasons why regression is such an important statistical method.

• Things to Remember •

1. Regression is the process of producing an equation.
2. At its simplest, this equation is in the form of $y = bx + c$.
3. y is what we want to predict or understand (such as how long it takes to run a marathon).
4. What we use to predict or understand y is called x (e.g., how old someone is).
5. y is sometimes called the *dependent* variable and x the *independent* variable; x can also be called a *predictor* or a *covariate*.
6. In the equation, b is called a coefficient; c is called the *constant* or *intercept*.
7. If there is only one type of x, this is called *univariate* regression.
8. If there are multiple x's (such as age, gender and training miles), this is called *multivariable* or *multiple* regression. Each x will have its own coefficient.
9. Linear regression is used when y is a continuous variable (such as time to run a marathon).
10. Logistic regression is used when y is a binary variable (such as whether a man has prostate cancer or not).

SEE ALSO: *Why your high school math teacher was right: How to draw a graph; My assistant turns up for work with shorter hair: About regression and confounding*

for Discussion

....

1. In reality, no one tries to predict marathon times in terms of age, gender and training miles. What would you use instead?

2. In a regression such as $y = b_1x_1 + b_2x_2 + c$, the c is called the intercept or constant. Why?

3. Why do you think that y is called the dependent variable and x's the independent variables?

4. In the marathon running example, the coefficient for "female" in the univariate analysis was 24, that is, women took 24 minutes longer to run the marathon. In the multivariable analysis, the coefficient was 23 minutes. How would you explain the difference between these two coefficients?

5. *For enthusiastic students only:* What is the log odds in logistic regression? What do statisticians tend to report instead of coefficients for logistic regression?

─────────

NOTE: See page 180 for answer sets.

My assistant turns up for work with shorter hair: About regression and confounding

Sparkling Conversation

Things I find myself saying: To a person with a cold—"Yes, there has been something going around." To a guy who loses his beard from one day to the next—"So you shaved." To someone with suddenly shorter hair—"Oh, you've had a haircut." The sad fact is, most of the time, I have nothing interesting to say at all.

On the other hand, the hair example does tell us something about regression. Let's imagine that I'd lined up a few hundred men and asked you to guess whether they had cut their hair in the last week or so. In some cases it would be obvious (like a guy with ratty long hair down his back) and, on balance, you'd guess that those with shorter hair would be more likely to have had a recent cut. On the whole though, your guesses wouldn't be that good: you wouldn't know whether someone with medium length hair had just had it cut back from being long or had grown it out from being short.

The reason I know that my assistant had his hair cut was because I knew that his hair had been longer the day before and, of course, your hair length on Tuesday is a very strong predictor of your hair length on Wednesday. This tells us that if the world doesn't match up to a prediction, and you think the prediction was a good one, then there is a good chance that something else is going on. Now remember that regression is about prediction: we try to predict a dependent variable y (such as your marathon time) in terms of one or more x's (such as how hard you train). So regression is useful if we have a hypothesis about that "something else going on" (like a hair cut).

A lobbyist explains why 2500 calories worth of burger, fries and a shake isn't fattening

Fast food generally contains a lot of fat (like a cheeseburger) and sugar (like milkshakes) and, as I understand it, eating a lot of fat and sugar tends to lead to weight gain. I have a data set in which about 2000 Americans were asked questions about their diet and exercise habits. Nearly two-thirds of the people in the survey ate at a fast food restaurant at least occasionally, and their rate of obesity was higher (21% vs. 15%; $p < 0.01$) than those who didn't eat fast food. However, before I even start to think what to do with my findings, I am visited by representatives of the American Association of Junk Food Lobbyists. The lobbyists claim that burgers have nothing to do with obesity; it is just that poorer, less educated people tend to eat junk food, as do men, and these groups don't work out as much and have worse dietary habits in general.

Here is a phrase you don't often read: the lobbyists are right (up to a point). When I run further analyses on the data set, I find that income, education, gender and exercise are all associated with obesity. For example, the rate of obesity was lower in survey respondents who exercised compared to those who did not (16% vs. 21%; $p < 0.01$). I also find that income, education, gender and exercise are all associated with junk food. Of those who ate junk food, 55% worked out—somewhat less than the 65% rate of exercise in those who avoided junk food ($p < 0.01$).

The lobbyists see these results and arrange a celebratory lunch (double guacamole bacon burger with large fries, big gulp soda on the side). But while they are away, I use a logistic regression equation to predict the expected probability of obesity for everyone in the survey on the basis of income, education, exercise habits and gender. The mean probability of obesity among those who ate junk food was 20% compared to 18% in those who didn't eat junk food. But the actual rates were 21% and 15%. Because the difference in obesity rates is bigger than we expected, this suggests that the relationship between junk food and obesity is not simply due to differences in other things, such as exercise, that affect weight. In other words, the world didn't turn out as we predicted, so something else must be going on.

Sleep deprivation makes me anxious

My kids have a theory: getting up early makes me anxious. Consider the fact that I am more stressed during the week (when I wake up early) than during the weekend (when, just occasionally, I get to sleep in). My own view is summarized in the following diagram:

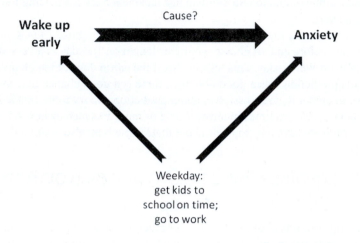

This is what statisticians called *confounding*—you think one thing causes another, but in fact it is something else entirely that causes both. It may not be that there is any real connection between early rising and anxiety, it is just that it being a weekday, and having work to do and kids to get to school, results in both me having to get up early and my anxiety level being higher. The ideal thing would be to find some times where I woke up early on the weekend or overslept during the week. We could then compare my anxiety level on early rising days and sleep-ins separately on weekdays and weekends. If, for example, I was more anxious on weekend days when I woke up early than on weekend days when I slept in, then we would have more faith that early rising really was associated with anxiety.

"Adjusting your results": Sounds somewhat naughty, but statisticians do it all the time

The problem with using a similar approach for our junk food data set is that we would have to compare obesity rates between junk food eaters and abstainers in a huge number of categories (wealthy, college-educated female exercisers, wealthy, college-educated female non-exercisers, not so wealthy, college-educated male exercisers, etc.) What multivariable regression does for us is to compare the effects of junk food in all the different groups all at once.

A multivariable regression for obesity gives y = log odds of obesity = $0.334 \times$ junk food $-$ $0.246 \times$ exercise $- 0.078 \times$ college educated $- 0.0858 \times$ income bracket $+ 0.375 \times$ male $- 1.27$ (see *When to visit Chicago: About linear and logistic regression*). In this regression, we call junk food the *predictor* (it is what we are really interested in) and exercise, education, income and gender the *covariates* (they help determine whether or not junk food is associated with obesity). As the coefficient for junk food is greater than 0, we conclude that eating junk food is associated with an increased risk of obesity. To work out whether this association is statistically significant we need the standard error, which is 0.121. The coefficient is nearly three times its standard error. Given that, under the null hypothesis, the coefficient will be within two standard errors of zero 95% of the time, we can conclude that there is indeed a statistically significant association between junk food and obesity (the *p*-value is actually 0.006).

You can think of the multivariable regression equation like this. Imagine that there were two groups of 100 Americans—one that ate junk food and one that avoided it—and that the two groups were exactly alike in terms of exercise, education, gender and income. If there were higher obesity rates in the junk food group, you'd want to say that this probably resulted from a true relationship between junk food and obesity because you couldn't use the lobbyists' argument that differences in other factors, such as exercise, were to blame. Now imagine the groups were the same other than that slightly more people in the junk food group didn't exercise. We wouldn't want to say, "Oh, the groups are a little bit different, we shouldn't even think about comparing them." What would be more sensible would be to say, "Ok, the groups are a bit different, so maybe we can adjust the results we see to make up for these differences."

As a simple example, we'll use the figures from our survey and say that the rates of obesity were 21% in the junk food group and 15% in the group that didn't eat junk food and the rates of exercise were 55% and 65% respectively. We are concerned that the lower rate of exercise in the junk food group might cause the differences in obesity rates. Let's say, for the sake of argument, that in analyzing our data we find that one in five of those who don't exercise are obese. There are 10% (i.e., 10) more non-exercisers in the junk food group, so you'd expect 20% of these, or 2, to be obese. We can now "adjust" our obesity rates in the junk food group. We imagine that, if the exercise rate in both groups was the same, there would be 2 fewer obese individuals in the junk food group. So instead of an obesity rate of 21%, the obesity rate would be 19%. This is still higher than 15%, the obesity rate among those who avoid junk food, so we conclude that the association between junk food and obesity isn't related to exercise.

In the multivariable regression, we make this sort of adjustment for exercise, income, education and gender simultaneously. Accordingly, we might report our results as follows: "After adjustment for exercise, income, education and gender, junk food was a statistically significant predictor of obesity (odds ratio 1.40; 95% confidence interval 1.10, 1.77; $p = 0.006$)."

A good example of how multivariable regression helps identify confounding is the link between income and crime (see the discussion section answers for *How to shoot a TV episode: Statistical analyses that don't provide meaningful numbers*). There are higher rates of violent crime in states with a higher median income, which suggests that a bunch of rich people are jumping out of their Maseratis to go rob banks and shoot people. But the relationship between income and crime is confounded by city living: people who live in cities tend to be wealthier, but there is also more crime in cities. If you adjust for city living in a multivariable model, increases in median income are associated with decreases in violent crime, which is what I guess you'd expect.

Roll up for the magic show!

Multivariable regression is mathematically very complicated. It is also now very easy to do because computers can run even the most complex regressions using pull-down menus. As a result, regression has become extremely popular and many non-statisticians seem to think that it has near magical properties. I have often seen scientists deflect quite reasonable criticisms of their work with "but we adjusted for that using multivariable regression" as if, by using regression, scientific problems miraculously disappear.

But regression remains an imperfect technique for several reasons. First, we can only adjust for what we measure. We don't always measure everything, and there are some things that cannot reasonably be measured. In the junk food data set, for instance, data were taken on exercise habits, but not on occupation. So someone who sat behind a desk five days a week and went for a brief jog on Saturday was described as an exerciser whereas someone employed in a profession that required demanding physical labor, but did not otherwise work out, was counted as not exercising. In theory, we could have asked about whether survey respondents had a strenuous job, but there are other things that would have been difficult to ask about. For example, culture is one of the strongest influences on diet, in that we tend to eat what our friends and family serve for dinner. Now I'd be fine with a survey interviewer asking me if I was Jewish, but I'd probably find it a little off if they then said, "Ok, but would you describe yourself as very Jewish, somewhat Jewish, or only a little bit Jewish?"

Even what we do measure we may not measure well. We can ask someone approximately how many times a week they work out, but they may not answer accurately. In fact, studies have shown that people tend to overestimate "good" behaviors ("Oh yes, I regularly eat healthy food.") and underestimate "bad" behaviors ("I do eat ice cream, but only rarely, and I only take a small serving."). Moreover, someone working out three times a week might be going for a gentle 2 mile jog or a fast 25 mile bike ride.

The third reason why multivariable regression is not a magic wand is because two predictor variables are often highly correlated and, in these cases, multivariable regression cannot tell them apart. As an obvious example, if every person who exercised avoided junk food and no one who ate junk food ever exercised, it would be impossible to tell whether differences in obesity were due to exercise or diet. In practice, this sort of problem is more subtle, but causes problems nonetheless.

The limitations of multivariable regression illustrate a general rule of statistics: statistics can't do for you what the science doesn't. Good statistics is a bit like a pair of high quality stereo speakers: it allows you to hear the data clearly without distortion; yet the best pair of speakers in the world isn't going to make a CD sound good if the music was badly played or recorded. If we don't have data, or the data are badly measured, or two things are so similar that it is difficult to tell them apart, then statistics can't help us no matter how complex the statistical techniques we use.

• **Things to Remember** •

1. We often think that one thing is associated with another (like getting up early and being anxious), when in fact both are caused by something else entirely (it being a weekday). This is known as confounding.

2. If the world doesn't turn out how you predict, and you think your prediction is a good one, it often means that something else is going on.

3. Regression makes predictions.

4. If you think that an *x* is associated with a *y*, but are worried about confounders, you can add these confounders as covariates in a multivariable regression.

5. You can calculate the statistical significance of a predictor *x* in a regression by comparing its coefficient to its standard error.

6. Multivariable regression is not magic, and it doesn't make the problem of confounding go away.

for Discussion

1. In our multivariable regression, junk food was associated with obesity even after controlling for income, gender, education and exercise. Can we conclude that eating junk food causes obesity?

2. I gave diet and exercise as an example of something that couldn't be measured precisely.

Couldn't we get people to complete a diary of everything they ate and all the exercise they did?

3. *For enthusiastic students only:* I reported a coefficient of 0.334 and a standard error of 0.121. How did I get the odds ratio of 1.40? How did I get the confidence interval?

NOTE: See page 184 for answer sets.

CHAPTER 20

I ignore my child's cough, my wife panics: About specificity and sensitivity

Two approaches to finding disease: Which is better?

My wife is extremely sensitive to our children's well-being: my son only has to scratch his head, or my daughter cough, and they find themselves being checked for lice and having their temperature taken. (Yes, that means *either* scratching or coughing will lead to *both* the lice comb and the thermometer.) I don't tend to check the temperature of a child with an itchy head. In fact, if I don't see clear evidence of double pneumonia, such as an x-ray with a follow-up MRI, then I can't see why everyone can't go to school as normal.

Imagine that, to resolve our dispute of who has a better approach to child health, my wife and I went around our neighborhood, checking on all the children. A doctor then examined all the children to make a final decision. The results of this experiment can be reported in a table:

		Doctor's diagnosis		
		Child really is sick	Child is fine	Total
My wife's assessment	Child might be sick	49	250	299
	Child is healthy	1	750	751
	Total	50	1000	1050

		Doctor's diagnosis		
		Child really is sick	Child is fine	Total
My assessment	Child might be sick	25	10	35
	Child is healthy	25	990	1015
	Total	50	1000	1050

You can see that my wife is good at working out whether a child is sick, whereas I am good at working out that a child is healthy. Roughly speaking, statisticians call this, respectively, *sensitivity* and *specificity.* In other words, my wife is sensitive and I am an old grouch (i.e., specific).

The *sensitivity* of a diagnostic test is defined as the probability of a positive test given that you are sick. Now in this case, a "positive test" is defined by whether or not my wife said the child was sick, but it could also be a blood test result from a laboratory, or a score on a question-naire (e.g., "a score of 8 or more counts as depression"). Sensitivity is calculated by counting the number of sick patients given a positive test result and then dividing by the total number of sick patients. There were a total of 50 sick children; 49 of these were picked out as sick by my wife. So my wife's sensitivity is 49 ÷ 50, or 98%; I thought that 25 of the sick children might be unwell and told the other 25 to stop whining: this gives a sensitivity of 25 ÷ 50 = 50%.

The *specificity* of a diagnostic test is defined as the probability of a negative test given that you are not sick. This is calculated by counting the number of healthy patients given a negative test result and then dividing by the total number of healthy patients. A total of 1000 children were healthy: I called 990 of these healthy for a specificity of 990 ÷ 1000 = 99%; my wife only thought that 750 of the healthy children were indeed healthy, a specificity of 750 ÷ 1000 = 75%.

Ok, so who has the better approach to child health?

If you are stumped, then don't worry. I posed exactly the same question to a group of profes-sional statisticians and they were equally unable to answer. This is sort of odd, because statisti-cians have spent years running around telling us to calculate sensitivity and specificity. Yet, as it turns out, sensitivity and specificity generally don't help you decide which of two tests is better.

The key point is that which approach is better, mine (low sensitivity, high specificity) or my wife's (high sensitivity, moderate specificity), depends on how important it is to find the disease and how harmful the treatment is if applied unnecessarily. Let's imagine that instead of looking for children with flu or head lice, we were trying to diagnose an infection, such as the black plague, that could be fatal if not treated with antibiotics. Missing a case of a disease would be a disaster—the patient might die—but treating someone with antibiotics who doesn't need it is not a huge problem. In other words, if there is a fatal infection going around, you'd be better off having my wife panicking than having me there telling you that you'll probably be fine, and that those large black boils covering your legs are just heat rash. On the other hand, if the treatment for a certain disease were surgery, you'd want to be pretty sure you had it before going under the knife, and would look for a diagnostic test with high specificity.

You're a doctor and the test has just come back positive. What do you tell your patient?

Remember that the definition of sensitivity is "the probability that the test will come up positive if you do indeed have disease." This would be extremely helpful if doctors went around saying to each other, "I have a patient here who is sick, guess what his test results were." ("Wait, wait, don't tell me, I'll look up the sensitivity of the test . . . ."). The problem that doctors face is that they have the test results and need to know whether the patient has the disease. Let's imagine that the company Genehype© sold a diagnostic test for strep throat. Here are the data that the company provided to the Food and Drug Administration to get the test approved:

		True diagnosis		
		Strep throat	Not strep throat	Total
Genehype© test	Positive	90	10	100
	Negative	90	110	200
	Total	180	120	300

The doctor wants to know the probability that the patient has a disease given a certain test result. You can see that 100 patients tested positive and 90 of these had strep throat. This means that a patient with a positive test has a 90% chance of really being sick and it would seem reasonable to go ahead and use antibiotics. There were 200 patients with a negative test result and 110 of these were free of infection. So a negative test means about a 55% chance of really not having strep throat. As a result, the doctor might say something like, "The test came back negative but that doesn't rule out strep throat, so I want you to keep an eye on things and get back to me if you don't get better in the next couple of days."

The statistics we just calculated are called the *positive predictive value* (defined as the probability you have the disease if your test is positive) and the *negative predictive value* (defined as the probability that you don't have the disease if your test is negative). You can see that they can actually help the doctor make a decision on what to do.

Is Martin's unresolved conflict with his mother affecting his marriage?

I was once chatting with a psychotherapist who told me about a case involving a young couple going through a bit of a crisis. The psychotherapist told me that the problem boiled down to the husband's mother. I don't know for sure whether he was hinting at something. (I get along pretty well with my mother.) Also, on the plus side, he didn't mention Oedipus. Nonetheless, I was rather put off by his apparent certainty: when I asked him about this, he said that he had "absolutely no doubt" that the issue boiled down to the mother.

This reminded me of a professor I met at college, a social worker who didn't see clients any more but spent her time teaching. Her specialty, she told me, was training other social workers to recognize signs of child abuse. I asked her how she knew she was right and that what she said indicated child abuse really did show that a child had been mistreated. Her response wasn't much different from the psychotherapist's, although she also volunteered that one shouldn't even raise the question because it would be unfair to the children.

I bring this up because issues of diagnosis—and therefore sensitivity, specificity and positive and negative predictive value—sound like they are just about medicine. But as it happens, "diagnosis" crops up in many other fields that use statistics, not just psychology and social work, but political science, sociology, and economics as well. In place of disease, we are interested in whether someone will vote, commit a crime or buy a product and instead of a diagnostic test, we might have information on a person's age, prior arrests or income. Nonetheless, we want to describe the relationship between the information we do have (e.g., the test result or a person's age) and the information we don't have (e.g., whether the person has a disease or whether they will commit a crime). Good statistics is about finding the best way to describe those relationships: whatever the psychotherapist and the social worker might imply, none of us is infallible.

• Things to Remember •

1. Sensitivity is the probability of a positive diagnostic test given that you have the disease.
2. Specificity is the probability of a negative diagnostic test given that you don't have the disease.
3. If you have two diagnostic tests and you want to know which one is better, sensitivity and specificity are often not that helpful.
4. To work out which of two diagnostic tests is better, you have to think about consequences: what would happen if someone had the disease but you called them healthy? What would happen if you mistakenly told someone that they did were sick and then gave them a treatment they didn't need?
5. If you have a patient in front of you and you want to know what to tell them, sensitivity and specificity are generally of little use: you need to know positive and negative predictive value.
6. You have a patient with you and the diagnostic test just came back positive. The probability that the patient really does have the disease is the positive predictive value.
7. You have a patient with you and the diagnostic test just came back negative. The probability that the patient is really free of disease is the negative predictive value.
8. It is kind of ok to panic about children's health.

SEE ALSO: *Avoid the sales: Statistics to help make decisions*

for Discussion

....

1. In the experiment to determine whether my wife or I had a better approach to child health, we compared our results to that of a doctor. I described the results of this experiment in terms such as: "There was a total of 50 sick children; 49 of these were picked out by my wife." To be more accurate, I should have said: "A total of 50 children were described by the doctor as sick; 49 of these were picked out by my wife." In other words, I am using "described by the doctor as sick" to mean "really was sick." But is the doctor always right?

2. Do sensitivity and specificity ever tell us which of two diagnostic tests is better?

NOTE: See page 186 for answer sets.

CHAPTER 21

Avoid the sales: Statistics to help make decisions

I once wandered into book store in Los Angeles looking for a novel. I came out with a cookbook that I didn't really want, couldn't really afford and was far too big and heavy to take back on the plane to New York. It was, however, 30% off and the thought of saving $40 had clearly put me into a sort of trance-like state, in which my hand reached for my credit card without my really being aware of it. So, whenever I visit my in-laws, I have an unopened 12 lb. tome to remind me that it doesn't really matter whether something is cheaper than it used to be. What matters is how much it costs right now and whether it is worth the price.

The connection between vacation shopping and my work life is that both Christmas sales and scientific papers about new cancer drugs advertise, say, 30% off: either a price that is lower by 30% or death rate decreased by 30%. But just as a 30% reduction in price doesn't tell us whether it is worth buying a book, a 30% reduction in death rate doesn't tell us whether it is worth using a drug. To understand why in mathematical terms, we need to think about a field of statistics known as *decision analysis*.

Imagine that you and I are playing a betting game where we both put $5 on the table and then one of us rolls a die: if it is a 4, 5 or 6 you win the money, otherwise I do. But then I make a proposition: if it is an even number or a 5, you pay me $10; if it is a 1, I pay you $2; if it is a 3, you roll again: then if on your second roll you get a 2 or a 4, I pay you $100; otherwise you get $5. Do you take my new bet or carry on playing as normal? Well of course the answer is to avoid the complicated bet, because I am a statistician, so I'd get these sorts of things right.

The following diagram shows why you'd be making a mistake to take my bet. Known as a *decision tree*, the diagram systematically identifies each possible outcome and then assigns a value and a probability to the outcome. You multiply the probability by the outcome to work out what you should expect to gain or lose for each of your two options, new bet or old bet.

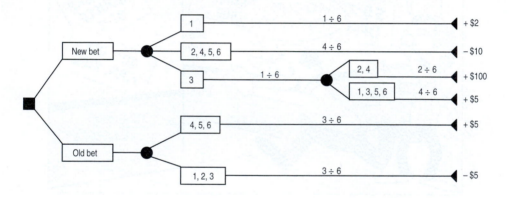

In the case of the complicated new bet, there are three possible outcomes: you win straight-away, lose straightaway, or roll again. If you roll again, you can either win a lot or a little. The probabilities and values of each of these outcomes are shown in the following table:

Outcome	Probability	Expected winnings
$2	$1 \div 6$	$1 \div 6 \times \$2 = \$0.33$
−$10	$4 \div 6$	$4 \div 6 \times -\$10 = -\$6.67$
$100	$(1 \div 6) \times (2 \div 6)$	$(1 \div 6) \times (2 \div 6) \times \$100 = \$5.56$
$5	$(1 \div 6) \times (4 \div 6)$	$(1 \div 6) \times (4 \div 6) \times \$5 = \$0.56$
Total	1	−$0.22

If you add up your expected winnings, you get that you expect to lose about 22 cents every time you play. If you decide to stick to our standard bet, you have a 50% chance of winning $5 and a 50% chance of losing $5. This comes out to an expected gain of $0—not great, but better than losing money.

Exactly the same principle can be applied to decisions about pretty much anything, although we'll stick with medicine as our example and take the case of bone marrow transplant for late stage cancer. In bone marrow transplant, patients are given very high doses of extremely toxic chemotherapy drugs. So toxic, in fact, that the drugs destroy the patient's immune system. To prevent the patient dying of an infection, doctors give patients new bone marrow cells; these grow and allow the patient to recover immunity. Nonetheless, treatment remains very dangerous and patients have a chance of dying from the treatment or having some complication such that treatment ends early with minimal benefit. If patients complete treatment, they have a chance of responding (meaning that the treatment kills most of the cancer cells), in which case they experience an important improvement in survival. The alternative to transplant is standard chemotherapy, to which patients have a chance of a limited improvement in survival but no risk of immediate death.

The following hypothetical decision tree is similar to that for the bet. Instead of dollars, the ends of the branches give average survival and instead of working out the probabilities from the dice rolls, they are obtained from estimates in the scientific literature.

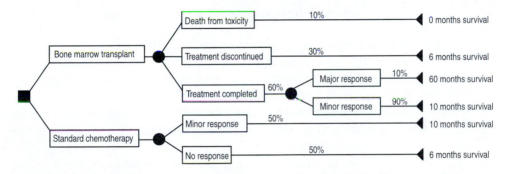

To find out whether a patient should consider an aggressive treatment, you do exactly the same math as for finding out whether you should take a particular bet: you multiply the probabilities by the outcome for each possible result and compare for each decision. In this (hypothetical) case, it turns out that expected survival is 10.8 months for bone marrow transplant and 8 months for standard chemotherapy.

What about our new drug, which decreases the risk of death by 30%? The obvious thing to notice is that there is nowhere in the decision tree to plug in this number. Statisticians describe "a 30% reduction in the risk of death" in terms of *relative risk*. For example, if in a drug trial the death rate at two years was 20% in the control group and 14% in the drug group, this would be a relative risk of 14% ÷ 20% = 70%, or a relative risk *reduction* of 30%.

Relative risks are well known to have little value for making decisions. If I told you that an expensive new bike lock halved the risk of your bike being stolen, you wouldn't know whether to buy it on that basis alone. You'd have to consider (a) the rate of bike theft in your town, (b) the cost of replacing your bike; and (c) the cost of the lock. Similarly, take the common news headline that something or other "increases your risk of cancer." This is only a worry if you are at reasonably high risk to start with. For example, one drug sometimes prescribed to young women can increase the risk of early breast cancer by 40 or 50%. But breast cancer in young women is fortunately very rare; in some women, an increase of 50% translates to only one breast cancer for every 10,000 taking the drug. This is the *absolute risk difference* and is far more useful for decision making.

The other interesting thing about the decision tree is the numbers given for "average" survival (e.g., 10 months survival, 6 months survival). These averages are means rather than medians (see *Bill Gates walks into a diner: On means and medians*). Means are better for decision making than medians. If you don't believe me, let's arrange the bet I first suggested in the discussion section of *Bill Gates walks into a diner*: if you roll a 1, 2, 3, 4 or 5, I pay you $20; if you roll a 6, you pay me $1,000. Your median winnings here are $20 and your mean winnings are negative $150. (This also explains why most of us avoid Russian roulette.) As it happens, in trials of new cancer drugs statisticians calculate medians rather than means. The reasons for this are mathematical, but the result is that many cancer trials have no immediate practical interpretation.

Statistics is the process of moving from an area of scientific study to math and then back to science. For example, we take a scientific question and turn it into a statistical hypothesis, or we take scientific data and calculate statistical estimates. We must then translate our hypothesis test, or our estimates, so that they mean something scientifically, or have some practical value. "Practical value" often means "helpful for making decisions" and this means absolute differences rather than relative differences, means rather than medians.

• Things to Remember •

1. Decision analysis is a type of statistical analysis that helps decision making.

2. The first stage of decision analysis involves four steps: (a) write down each possible decision, (b) work out all possible outcomes of each decision, (c) work out the probability of each outcome and (d) work out the value of each outcome.

3. A *decision tree* is way of showing all this in a picture.

4. The expected result of each decision is calculated by multiplying the probability of each outcome by its value and adding together.

5. If the risk of something is normally r and changes to s, the relative risk is $s \div r$ and the absolute risk difference is $s - r$. For example, if the risk of a manufacturing error is normally 2% (r), and this increases to 3% (s) on hot days, the relative risk is $s \div r = 3\% \div 2\% = 150\%$ and the absolute risk difference is $s - r = 3\% - 2\% = 1\%$.

6. Relative differences are not useful for decision making, in the same way that a sale item might not be worth buying even if it is 30% cheaper than normal.

for Discussion

1. Using a regression equation, Helen's doctor calculates that her risk of a heart attack is 8%. She is told that if she takes a cholesterol lowering drug, her risk will be reduced by 25%. However, the drug raises her risk of cancer by 0.5%. How much does Helen's risk of a heart attack decrease in absolute terms if she takes the drug? Do you think she should take it? What if her risk of heart disease was 2%?

2. I gave an example of decision analysis for a medical decision. However, decision analysis did not develop in medicine, but in another field of statistics. Which?

3. Decision analysis isn't widely used. Why do you think not?

NOTE: See page 187 for answer sets.

One better than Tommy John: Four statistical errors, some of which are totally trivial, but all of which matter a great deal

Tommy John, the renowned pitcher, once made three errors on a single play: he bobbled a grounder, threw wildly past first base, then cut off the relay throw from right field and threw past the catcher. I was reminded of that story when reading a scientific paper describing a clinical trial comparing a new pain drug with a placebo. Near the start of the results section, the authors wrote something like, "Although there was no difference in baseline age between groups ($p = 0.458$), controls were significantly more likely to be male ($p = 0.000$)."

This actually goes one better than Tommy John, because there are actually four errors in this single sentence (or perhaps even four-and-a-half).

1. **Accepting the null hypothesis.** You cannot conclude "no difference" between groups on the basis of a high *p*-value. This is because failing to prove a difference is not the same as proving no difference. When we conduct a statistical test, we state a null hypothesis (e.g., "There is no difference between groups"), then calculate a *p*-value. If the *p*-value is statistically significant, we reject the null hypothesis. If the *p*-value is non-significant it isn't that we accept the null hypothesis, we *fail to reject* the null hypothesis (see *Michael Jordan won't accept the null hypothesis: How to interpret high p-values*).

2. **Giving *p*-values for baseline differences between randomized groups.** The way that trials of drugs are conducted is that the researchers chose at random which patients receive the drug and which receive the placebo. Such studies are known as *randomized controlled trials*. The idea behind randomization is to make the groups as similar as possible so that any differences at the end of the study can be attributed to the drug. Baseline differences at the beginning of the trial, such as in age or gender, are due to chance, just as I might flip a coin 20 times but not get exactly 10 heads and 10 tails. We use *p*-values to test hypotheses, in this case, a null hypothesis that can be informally stated as: "There is no real difference between groups; any differences we see are due to chance alone." In short, giving a *p*-value for baseline difference between groups created by randomization is testing a null hypothesis that we know to be true (I also discuss this point in *Choosing a route to cycle home: What p-values do for us*).

3. **Inappropriate levels of precision.** The first *p*-value in our multi-error sentence is reported to three significant figures ($p = 0.458$). What do the 5 and 8 tell us here? We are already way above statistical significance. A little bit more or less isn't going to change our conclusions, so reporting the *p*-value to a single significant figure (i.e., $p = 0.5$) is fine. Inappropriate levels of precision are ubiquitous in the scientific literature, perhaps because a very precise number sounds more "scientific." One of my favorite examples is a paper that reported a mean length of pregnancy of 32.833 weeks, suggesting that we want to know the time of conception to the nearest 10 minutes. This would require some rather close questioning of the pregnant couple.

4. **Reporting a *p*-value of zero.** No experimental result has a zero probability. Even if I throw a thousand dice I have a small, but definitely non-zero, chance of getting all sixes. I once pointed this out to some researchers who had cited a zero *p*-value in a paper, only to have them reply that the statistical software had given them $p = 0.000$, so the value must be right.

This gets to the heart of why I care about these errors even though they don't make much difference to anything. (Why don't I just ignore those unnecessary decimal places?) Many people seem to think that we statisticians spend most of our time doing calculations, but that is perhaps the least interesting thing we do. Far more important is that we spend time looking at numbers and thinking through what they mean. If I see any number in a scientific report that is meaningless—a *p*-value for baseline differences in a randomized trial, say, or a sixth significant figure—I know that the authors are not being careful about what they are doing, they are just pulling numbers from a computer printout. Statistics is more than just cutting and pasting from one software package to another. We have to think about what the numbers mean and the implications for our scientific question.

• **Things to Remember** •

1. Don't accept the null hypothesis. Instead say something like "we could not show a difference."
2. Don't report *p*-values for baseline differences between groups created at random. Simply report estimates for each group separately.
3. Think carefully about the number of decimal places you report.
4. Don't give a *p*-value of zero; say something like "$p < 0.001$" instead.

for Discussion

1. I mentioned that the sentence had "half an error." What was it?

NOTE: See page 189 for answer sets.

Weed control for *p*-values: A single scientific question should be addressed by a single statistical test

My garden has set several weed world records, in both the size and variety categories. Now a few weeds are neither here nor there, but when the biomass of weeds reaches a certain critical point, they choke out the stuff you are meant to look at (like the grass and the flowers). You can say something similar about many scientific papers—*p*-values growing like weeds, choking the science until you can't tell what you are meant to be looking at.

In a typical laboratory investigation, groups of mice are injected with salt water, or one of several increasing doses of a drug or toxin, and then some measurement is made, say, of immune function. The results are generally presented in a bar chart, such as the following. The stars indicate the doses for which there were statistically significant differences in immune score compared to control.

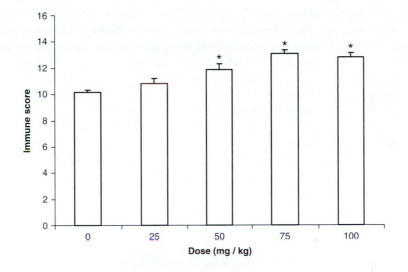

The researchers might then give *p*-values for comparisons between all doses, reporting, for example, that 100 and 75 but not 50 mg/kg were statistically superior to 25 mg/kg. This could be taken to suggest that although 50 is effective and 25 isn't, there is no difference between 50 and 25.

The problem here is twofold. First, the researchers are testing multiple hypotheses instead of just one. The design of the experiment implies a single question (What is the relationship between dose and immune function?) rather than several different questions (Is 75 better than 25? What about 50?). Second, the researchers are treating each hypothesis as independent: when comparing two doses, they proceed as if no other data exist. This goes against scientific common sense. If I told you that 25 mg/kg and 75 mg/kg both led to higher immune scores than control, you would probably feel comfortable betting that 50 mg/kg also improved immunity. By analogy, if at the end of the season Tampa Bay had won more games than the Red Sox, and the Red Sox won more games than the Yankees, you wouldn't go on to ask whether Tampa Bay had won more games than the Yankees.

A better approach to these data provides a nice illustration of the principle that a single scientific question should be converted into a single statistical hypothesis, which is then tested by a single *p*-value. What I would do is to create a linear regression (see *When to visit Chicago: About linear and logistic regression*). This would produce the equation $y = bx + c$, where y is the immune score and x is the dose. Some typical data would give y = immune score = 0.030 × dose + 10.2, meaning that immune scores increase by an average of 0.030 for a 1 mg/kg increase in the dose of the drug. I also find that the coefficient a (0.030) is nearly seven times larger than its standard error (0.0044) so I also get a very low *p*-value. We can conclude that we would be unlikely to get data like this if the coefficient was really zero, so we should reject the hypothesis of no association between the dose of the drug and immune function in mice.

In the meantime, I need to work out whether 500 mg of aspirin is any better than 499 mg for treating headaches caused by a surfeit of *p*-values. Also, I need to complete my letter to the president pointing out that a tax rate of 24.9% compared to 25% will have no statistically significant difference on the national debt, so please could I pay less tax? And clearly my children will never get any taller, because there is no statistically significant difference in their height from one day to the next.

If you get my point, then you are doing quite a bit better than many practicing scientists. One time, I received a scientific paper for comment from a young pathologist, who had analyzed data from a series of 150 or so patients with cancer. Halfway through the results, the *p*-values started to swim before my eyes: there were *p*-values for different types of cancer, in different locations, separately by age, with all analyses then repeated depending on how the cancer was found. I ended up counting 126 separate *p*-values—that is, nearly one scientific question for each patient in the study. It sounds ridiculous when you put it like that, but it is all too easy to generate endless lists of *p*-values using statistical software, regardless of whether any of them address a question you actually want to answer.

Which is to say, just as I go through my garden on a Saturday, pulling out the weeds, you have to weed out excess *p*-values from a scientific report. Otherwise, well, you just won't be able to see the flowers.

• Things to Remember •

1. Each and every *p*-value you report addresses a specific null hypothesis and therefore a specific scientific question.

2. Questions about how things change over time, or vary with dose, can be addressed by a single regression equation, instead of multiple comparisons between selected pairs of doses or times.

> ❖ **SEE ALSO:** *Boy meets girl, girl rejects boy, boy starts multiple testing.*

for Discussion

1. What does our regression assume about the association between the dose of the drug and immune function?

2. The authors of the immunity study found no significant difference between the 50 and 25 mg/kg doses. Does this really mean that "there is no difference between 50 and 25"?

NOTE: See page 189 for answer sets.

How to shoot a TV episode: Statistical analyses that don't provide meaningful numbers

A friend of mine is a cameraman on a TV series about cops and robbers. It sounds like an interesting job, so I asked him one day how it all worked.

· · · · · · · · ·

Me:	*What do you do first in the morning?*
Cameraman:	*When we get to the location we spend about 2.8 setting up.*
Me:	*2.8 hours?*
Cameraman:	*No, just 2.8. Anyway, then we have to call the actors and as many as _____ of them have emergency appointments with their divorce lawyer, or are having an artistic crisis or something, and can't be found.*
Me:	*Sorry, I didn't catch that. You've had how many actors go missing?*
Cameraman:	*It is impossible to work out.*
Me:	*Can't you just count?*
Cameraman:	*No. Anyway, making a show sounds like fun, but the hours are long, and working with TV types can be a hassle. The pay is good, though: I can get 0.9 for a day's filming.*
Me:	*How much is that?*
Cameraman:	*More than 0.8.*

· · · · · · · · ·

This conversation reminded me of three statistical methods that are among those I worry about the most: correlation, ANOVA and chi-squared. In my view, these methods are often used without sufficient awareness that the numbers they provide are sometimes less than fully meaningful.

Here are some things you may have been taught about correlation:

- Correlation measures the strength of association between two continuous variables, like height and weight, or a country's wealth and its fertility rate.
- Correlation is measured from -1 to 1.
- A correlation above 0 means that as one variable increases, so does the other (like calories and weight). This is called a positive correlation.
- A correlation less than 0 means that as one variable increases, the other decreases (like exercise and weight). This is called a negative correlation.
- Correlations close to 0 are called weak; correlations close to 1 or -1 are called strong.
- Two things might be correlated, but that doesn't mean one causes the other, like stork populations in Europe and the birth rate in Germany (see discussion answers for *My assistant turns up for work with shorter hair: About regression and confounding*).

Here is something you probably haven't been taught about correlation: what it actually means. Take, for example, the correlation between a state's rate of poverty and its violent crime rate, which is about 0.27. Alright, because the correlation is above zero we know that states with a high poverty rate also tend to have high crime rates. Also, the correlation isn't that close to 1, so it is not as if only states with a lot of poverty have a lot of crime. But beyond those vague generalities, it is somewhat hard to say more about a correlation of 0.27.

Chi-squared has a similar problem. Here are some data from a survey that asked whether it is ok for people from different religions to get married:

Question 6: Attitude to interfaith marriage	Question 12: Proportion of friends of the same religion			
	All	**Most**	**Half**	**Less than half**
Mind a lot	36 (21%)	34 (6%)	8 (3%)	2 (2%)
Mind a little	39 (22%)	109 (18%)	24 (8%)	8 (8%)
Do not mind at all	99 (57%)	461 (76%)	276 (90%)	87 (90%)

The results are not enormously surprising: leaving aside whether this is right or wrong, people who only spend time with members of their own religion are less comfortable with interfaith marriages. Chi-squared here is 102.7, giving a p-value <0.0001. So what do you conclude? The null hypothesis is that there is no difference in attitude to interfaith marriage in terms of the proportion of friends with the same religion. The p-value is low, so you reject the null hypothesis. To show why this isn't an interesting conclusion, have a look at this table:

Question 6: Attitude to interfaith marriage	Question 12: Proportion of friends of the same religion			
	All	**Most**	**Half**	**Less than half**
Mind a lot	8 (3%)	2 (2%)	36 (21%)	34 (6%)
Mind a little	24 (8%)	8 (8%)	39 (22%)	109 (18%)
Do not mind at all	276 (90%)	87 (90%)	99 (57%)	461 (76%)

What I did was swap some of the columns so that it is those who count about half of their friends as being the same religion who mind most about interfaith marriage. Yet the chi-squared value is identical: 102.7. In other words, chi-squared gives you exactly the same result whether having mixed religion friends is associated with higher tolerance for interfaith marriage, a lower tolerance, or whether tolerance starts high, goes down and then goes up again.

The authors of the study used chi-squared because they were studying the relationship between two categorical variables (that is, variables that can only take a limited number of values, such as "all", "most", "half" or "less than half"). To examine the association between a categorical variable and a continuous variable (one that can take many values, such as weight or blood pressure), statisticians often use analysis of variance (ANOVA). ANOVA has become particularly popular for analyzing the results of experiments where several treatments are compared. As an example, imagine that you were interested in psychotherapy interventions for children with behavioral problems, where the extent of the behavior problem was measured on a 0–20 scale. If you had scores from children who either had or had not received an intervention (say, one-on-one psychotherapy), you could compare these scores by t test. However, let's imagine that you were interested in two different interventions (one-on-one psychotherapy versus one-on-one psychotherapy plus family psychotherapy). Now you have three groups (two different treatments plus a group with no treatment), and you can't use a *t* test to compare three groups. ANOVA allows you to throw all the data into one analysis and get a *p*-value.

If the *p*-value from ANOVA was less than 0.05, you would reject the null hypothesis of no difference between groups, leading to the conclusion that "there are differences in behavior depending on treatment." This is a pretty uninteresting conclusion, all told: it could be, for example, that psychotherapy makes things worse (see, for example, the "scared straight" study in *The probability of a dry toothbrush: What is a p-value anyway?*); it might also be that only one type of therapy is effective.

Here are some example results that would give a statistically significant *p*-value from ANOVA:

Group (30 children in each group)	Mean behavior score	Standard deviation of behavior score
No treatment	12.6	1.76
Psychotherapy	12.0	1.75
Psychotherapy plus family therapy	10.1	2.02

These data seem to suggest that one-on-one child psychotherapy may help a bit, but it is really the family therapy that is doing the work. This would make sense if you believed that what helps children most is day-to-day interactions with their family rather than one hour a week with a therapist. Either way, this is a far more useful presentation of the data than "the three groups are not the same."

Correlation, chi-squared and ANOVA are a good way of thinking about the difference between inference and estimation (see *I tell a friend that my job is more fun than you'd think: What is statistics?*). Chi-squared and ANOVA are limited because although they provide a *p*-value to test a hypothesis about an effect, they don't provide any idea of how large or small the effect is, or even in which direction the effect goes. In other words, chi-squared and ANOVA provide inference but not estimation. On the other hand, correlation does provide an estimate, but it isn't a useful one: we know that the association between crime and poverty is "0.27" but don't really get a good idea of what "0.27" means.

This suggests that we might sometimes need to complement correlation, chi-squared and ANOVA with methods that give additional numbers. In the case of the psychotherapy data set, we could use a *t* test to compare psychotherapy with no treatment, and then psychotherapy plus family therapy with psychotherapy alone. This analysis suggests that psychotherapy is associated with a 0.6 point improvement in behavior scores (95% confidence interval −0.3, 1.5) and that family therapy is associated with an additional 1.9 point improvement (95% confidence interval 0.9, 2.9).

As for the marriage data, probably the easiest thing to do is simply to combine the categories so that you have just two groups. For example, you could define "number of friends of the same religion" as "all" or "most" vs. "half" or "less than half," and define "attitude to interfaith marriage" as "mind a lot" or "mind a little" vs. "not mind at all." We would conclude that 28% of those with all or most of their friends being the same religion had an issue with interfaith marriage compared to only 10% of those who counted half or more of their friends as being in a different religion (a difference of 18%; 95% confidence interval 13%, 22%). With respect to the crime data, we could use linear regression to calculate for a 1% increase in the poverty rate—the number of violent crimes increases by 17 per 100,000. This is a number far easier to understand in real terms than a correlation of 0.27.

You may have seen tables or flow charts in textbooks that describe the type of statistical analysis you should use depending on the type of data you have. For example, such a table might tell you categorical data should be analyzed by chi-squared test, whereas a two-group comparison of a continuous outcome should be by Wilcoxon or t test. This is a good start, and I'd hardly want you trying to analyze data from the psychotherapy experiment using chi-squared. But here is a simpler rule of thumb: whatever analyses you do, test hypotheses that are interesting and provide numbers that mean something.

• Things to Remember •

1. Statistical analyses provide us with numbers.

2. The value of any statistical analysis depends on whether the numbers it provides are meaningful.

3. Many statistical methods, such as chi-squared and ANOVA, only provide p-values. As such, they can help test hypotheses (inference) but do not provide estimates.

4. We generally need to know how large or small something is in order to assess its importance. Accordingly, we generally like our statistical methods to provide estimates as well as inferences.

5. Some statistical methods, such as correlation, provide estimates that do not have an obvious meaning.

6. Chi-squared, ANOVA and correlation have their uses, but they often need to be complemented by other methods. For example, methods that provide estimates can be used alongside chi-squared and ANOVA.

for Discussion

1. I suggested that, for the crime data set, regression would give us a more meaningful number than correlation. Another alternative involves no numbers at all. How might you investigate these data without reporting specific numbers?

2. Should you never use chi-squared, ANOVA or correlation?

3. Is correlation really a dimensionless number, like it taking "2.8" to set up a film shoot?

4. Are we interested in inference for the crime data set? Should we report p-values from our analyses? What about confidence intervals?

NOTE: See page 190 for answer sets.

Sam, 93 years old, 700 pound Florida super-granddad: Two common errors in regression

I'd like to introduce you to Sam. Sam is a 93-year-old retiree living in Florida. He fought in the Pacific during the second World War, went to college on the GI Bill, then married his high school sweetheart and settled down to raise a family in Levittown, New York. He retired, moved to Florida, lost his wife to cancer and is now enjoying what his daughter calls "a second bachelorhood" with the numerous widows of Fort Lauderdale. The thing you need to know about Sam is not that he goes to the local diner at 5 every night for the "early bird special," nor that he is now so stooped over that he needs three cushions to see over the dashboard of his Chevy. The interesting thing about Sam is that he weighs almost 700 pounds, but can jump over an articulated truck and bench press nearly half a ton. Also, he is such a fast runner that he can sprint to the beach, a mile from his condo, and arrive 5 minutes before he sets out.

Sam is not a figment of my imagination, he is a statistical error. Here is Sam's data set:

Age	Weight (lbs)	Time for the one mile run (minutes:seconds)	High jump	Bench press (lbs)
2	30			
5	46			
10	75			
12	84	5:40		
14	114	5:05		
16	136	4:40	6ft 2 in.	160
18	143	4:35	6ft 6 in.	180
20	155	4:30		

As you can see, Sam was a pretty good athlete in high school, and kept up the running after he left. So how did he get to weigh 700 pounds? We can graph out Sam's weight against his age:

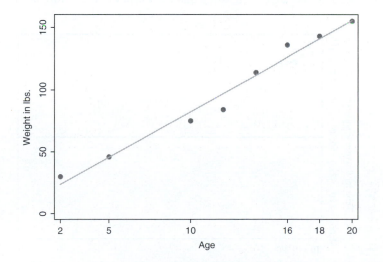

Each dot gives Sam's weight at a particular age and the line is the regression line that best fits the dots. The formula of the line is $y = 7.33 x + 9$. As y is weight and x is age, this gives weight = $7.33 \times$ age $+ 9$. You can see how we get to Sam weighing nearly 700 pounds at age 93: $7.33 \times 93 + 9 = 691$ lbs. This can also be shown on the following graph:

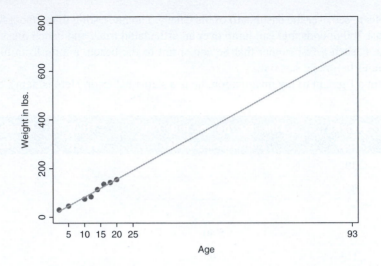

What you can see here clearly is that all the data are off to the left, and we are extending the regression line way out to the right where there are no data at all. We have no idea what happened to Sam's weight between 20 and 93, and certainly can't assume that it will have the same relationship to age as when Sam was a child. A statistician would say that we are *extrapolating* too far from the data.

The other obvious point about the figure is that it is a straight line. Sam's running times provide a good example of where a straight line makes no sense at all, because Sam's time for the mile eventually becomes negative, suggesting that he finishes a race before he starts.

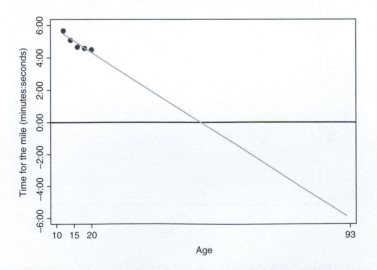

If you look carefully at the top left of the figure above, you can see that Sam's times don't even seem to start on a straight line—the regression line only goes through two of the five points, which seem to lie on a curve. As it happens, most things in life seem to follow a curve. Last night I cooked a stew, and added a couple of cloves of garlic. The reason I didn't add more than a couple is because the relationship between flavor and amount of garlic is a curve, not a line. Adding a little bit of garlic adds flavor; add too much and you ruin your food. (Not that the pasta place on 5th Avenue seems to notice.)

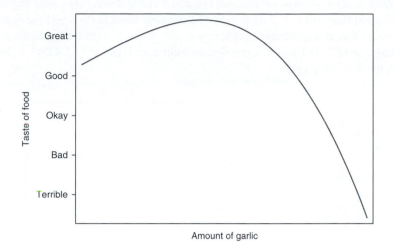

Or take the "law of diminishing returns." It takes me about 10 minutes after dinner to clear the plates from the table and the serving dishes from the counter, after which the state of the kitchen improves from "awful" to "not a disaster"; to actually wash the dishes and wipe down the surfaces—improving the kitchen from "reasonable" to "clean"—takes a good 30–45 minutes; to make the kitchen spotless (e.g., cleaning the oven, washing the pantry doors) would take all night.

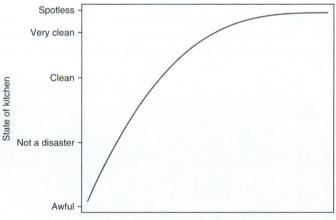

The law of diminishing returns can be defined as "the increase in reward per unit effort decreases with increasing effort." If y is reward and x is effort, this means "the increase in y when x increases by 1 decreases as x gets higher and higher." A regression line such as $y = 7.33\,x + 9$ has the same increase in y for when x increases by 1 for all values of x, and so won't fit the law of diminishing returns.

A straight line doesn't work for the relationship between running time and age, garlic and flavor or kitchen cleanliness and time spent cleaning. What you need instead is a curve and what you may remember from high school (or even the chapter *Why your high school math teacher was right: How to draw a graph*) is that you can plot a curve if you make your regression a little more complicated than just $y = 7.33\,x + 9$. The simplest thing to do is add the square of x. You might remember this as a quadratic equation: $y = ax^2 + bx + c$ (indeed, statisticians often call x^2 a "quadratic term"). If I run a regression including x^2 I get: $y = 1.429x^2 - 54.2x + 784$, which gives the following graph:

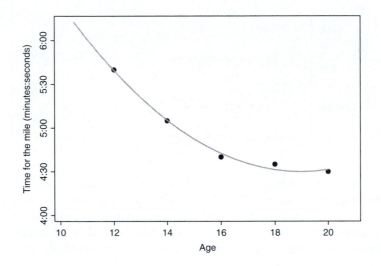

You can see how adding in x^2 gives you a curve that fits the data (i.e., joins the dots) better than a line. It would be easy to get comfortable at this point, and go home for a well earned dinner. But can we predict Sam's time for the mile any better? Here is what we get if we extend the regression line to 40:

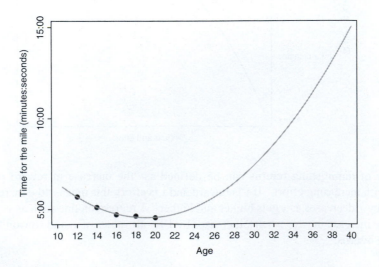

It is hard to believe that once he hit 40, an ex-high school athlete like Sam would take nearly 15 minutes to run a mile—not much faster than walking pace. This goes to show that you can make a regression as fancy as you want, but if you don't have the data, you may end up saying something just as ridiculous as having a 93-year-old bench press half a ton.

• Things to Remember •

1. The most common approach to regression ($y = bx + c$) assumes that the relationship between the predictor variable (x) and the dependent variable (y) follows a straight line.

2. Many associations follow a curve rather than a straight line. The law of diminishing returns is one example.

3. Regression can be used to describe associations that follow a curve. A typical approach is to use a quadratic term (x^2) for the regression equation $y = ax^2 + bx + c$. However, there are other techniques that statisticians use.

4. It is a bad idea to use regression to predict y for values of x far from any of those on your data set.

---

for Discussion

1. The regression for Sam's mile time was based on five data points, his time at ages 12, 14, 16, 18 and 20. Any thoughts as to whether this is a good or bad idea?

2. How did I work out that Sam could bench press a half a ton at age 93?

3. *For enthusiastic students only:* In the regression equation where y was time for the mile in seconds and x was age, I gave $y = 1.429x^2 - 54.2x + 784$. Why is the coefficient for x^2 (1.429) given to three decimal places whereas only a single decimal place is given for the coefficient for x (54.2)?

NOTE: See page 194 for answer sets.

Regression to the Mike: A statistical explanation of why an eligible friend of mine is still single

Mike is nice looking, has a great sense of humor, a good job and a fabulous apartment in a popular part of town. He is, however, still single, and he thinks the apartment is to blame. Over the past few years he has rented out the second bedroom to a variety of single friends and, without fail, all have succumbed to what he terms "The Curse of Mike": they meet someone, get married and move out.

Mike has come to believe that the second bedroom has some kind of magic romantic properties, so much so that he has considered moving into it. His friends think that his single condition is related to a repressed resistance to commitment buried deep in his subconscious. His mother says, naturally,

"Who would be good enough for my perfect bubula?" My explanation, of course, is statistical: in its general form it is known as regression towards the mean. Take any phenomenon of note at all—a guy in his 30's being single; Mr Jones' unexpectedly high blood pressure; Mrs Jones' arthritis having a flare-up; it being a particularly warm January—then wait a while, go back and take a look. Chances are that things will have returned to a more average stage: the man got married; the Joneses are feeling better; and it was below freezing for most of February.

Here is a simple illustration. Have a class of students roll a die, ask anyone who rolled a 1, 2 or 3 to leave the room and have the remainder roll again. In most cases, the second die roll will be lower than the first and overall, the mean of the second die roll will also be lower. For example, you might see something like this:

Student	First die roll	Second die roll	Second die roll lower?
Laura	4	5	No
Evangeline	4	3	Yes
Dena	3	—	
Eric	5	5	No
Felicia	1	—	
Aviv	6	5	Yes
Carlos	5	2	Yes
Michael	6	2	Yes
Madison	3	—	
Lev	2	—	

Six students—Laura, Evangeline, Eric, Aviv, Carlos and Michael—are still in the class for the second die roll. The mean of their first die roll is $4 + 4 + 5 + 6 + 5 + 6 = 30 \div 6 = 5$; the mean of their second die roll is $5 + 3 + 5 + 5 + 2 + 2 = 22 \div 6 = 3^2/3$. The true mean of a die roll is $3^1/2$; $3^2/3$ is closer to $3^1/2$ than is 5, so we have regression to the mean.

Although this seems trivial, it is exactly what we do in many experiments: take some measurement and measure again only if scores are high. For example, in medical trials, we measure patients' pain, blood pressure or anxiety, exclude anyone with low scores (on the grounds that they are not in need of treatment), give some treatment and then measure again after a few weeks. Because of regression to the mean, we'd expect just by chance that many patients would get better and that mean pain, blood pressure or anxiety would fall, even if treatment was ineffective. Or take a business study: a manager at the central office of a bank might identify a couple of branches with particularly long waiting times and test a change in procedure designed to shorten the line. Again, because only bank branches with long wait times were selected, we'd expect waiting times to fall by regression to the mean. This is just one of the reasons why it is so important to have control groups in experiments.

Perhaps my favorite example of regression to the mean is the "Curse of *Sports Illustrated*." The so called "curse" is based on the observation that athletes making the cover of *Sports Illustrated* typically have a rapid decline in performance, or get injured, shortly after being featured. Of course, the reason why athletes get picked out to be put on the cover is that they have done something spectacular ("Pedro's Amazing April," etc.) and at any randomly picked subsequent time are likely to be just average (Pedro gives up a two-run shot in the bottom of the ninth). Another sporting regression to the mean is that most coaches are confident in their ability to help an athlete to bounce back after a particularly poor performance. Again, this is because no matter what you do—whether you scream, comfort or threaten—the athlete is likely to have an average (i.e., better than poor) game next time out.

Regression to the mean pops up in just about all areas of statistics. So whether you are trying relieve pain, improve customer service or win at sports (or just meet someone nice), it is worth bearing in mind that things tend to average out in the end.

• Things to Remember •

1. Things vary around their mean.
2. If you see something that is far from the mean, it is likely to be closer to the mean the next time you check. This is called regression to the mean.
3. I will not respond to requests for Mike's phone number.

Discussion

1. Does regression to the mean explain why Mike is single?
2. What is the connection between regression to the mean and linear or logistic regression, the statistical technique used by statisticians to quantify relationships between variables?

NOTE: See page 195 for answer sets.

CHAPTER 27

OJ Simpson, Sally Clark, George and me: About conditional probability

I DON'T CARE WHAT THE LAB REPORT SAYS, THIS GUY DOESN'T HAVE HEART DISEASE!

The cast

1. OJ Simpson was a star football player whose subsequent Hollywood career was cut short when he was accused of killing his wife, Nicole Brown. He spent an afternoon driving around Los Angeles pressing a gun against his head while being chased in slow motion by the police and then turned himself in. The subsequent court case was about all anyone got to hear about for the next year or so. He was finally declared not guilty, although about ten years later he robbed a collector of sports memorabilia and was sent to jail.

2. Sally Clark was a British woman who lost two babies to "sudden infant death syndrome." Just as it sounds, this is when an apparently healthy baby dies for no apparent reason (it is sometimes called "crib death"). Someone clearly decided that Sally Clark had not suffered enough, because she was accused of murdering her babies, found guilty and sent to jail. She was later released after a legal appeal.

3. George is an incredibly athletic guy in his 20s who burned me deep several times playing Ultimate (a sport played with a Frisbee).

4. Me: I am no longer in my 20s. Also, I had trouble bending my knee the day after the Ultimate game. On the plus side, I do have a basic grasp of math.

Act 1: Statistics on national TV

The OJ Simpson trial was televised live and attracted massive ratings. As a result, millions of Americans got to watch OJ's lawyers use statistical reasoning to defend their client. One of OJ's main problems was that he had a history of physical violence against Nicole Brown. The prosecution argued that once a man had hit and punched his wife, it wasn't much of a stretch to imagine that he might, if provoked, go one step further and stab her. The lawyers' rebuttal went roughly like this:

- Statistics show that about 1 in 20 married men have hit their wives.
- Statistics also show that only about 1 in 20,000 men kill their wives.
- Dividing one number by the other you get that only 1 in 1000 men who hit their wives go on to kill them.
- The fact that OJ Simpson hit Nicole Brown gives him only a tiny chance of murdering her.

The problem with this argument—and the prosecution, being non-statisticians, didn't bring this up—is that Nicole Brown was already dead. It might well be true than a man who beats his wife probably won't turn to murder, but the question isn't "If a man beats his wife, what is the chance he will subsequently kill her?" What we want to know is "If a woman has been murdered, and she had previously been beaten by her husband, what is the chance that he was the perpetrator?" It turns out that if you look at the data on murdered women, then focus on those who had previously reported domestic violence, most of those were indeed killed by their husbands.

Act 2: Statistics by a pediatrician

At the trial of Sally Clark, Dr. Roy Meadows, a pediatrician, tried to emulate OJ's lawyers' attempts at amateur statistical analysis. (Question to Dr. Meadows: would you like me to treat a sick relative of yours?) Meadows first showed data that only about 1 in 8500 babies die of crib death. He then asserted that the chance of two crib deaths was 1 in 8500 multiplied by 1 in 8500, what was reported in the press to be "1 in 73 million." This is fantastically unlikely, therefore the deaths didn't happen by chance, and Sally Clark was probably responsible.

Meadows used the same reasoning you'd use to calculate the probability of coin tosses: the chance of throwing heads is 0.5, so the chance of throwing two coins and getting both heads is 0.5×0.5, and the chance of three heads in three tosses is $0.5 \times 0.5 \times 0.5$. But there is a big difference between coin tosses and illness in children—what statisticians call *independence*. Two coin tosses are independent because the result of the first toss gives you no information about the second: if I throw heads and ask you to give the probability that the next throw also comes up heads, you'll answer "0.5," exactly the same number you'd have given if I hadn't told you about the previous coin toss. But if I tell you that a woman has a sick child, you'd probably guess that her second child had an above average chance of being sick because many diseases tend to run in families. The probability that one child is sick is therefore not independent of his or her sibling's probability, and you can't just multiply probabilities together to work out the chance that both siblings are unwell.

But let's assume that Meadows was right, and that the probability that Sally Clark would lose two children to crib death was indeed 1 in 73 million. We still have a problem very similar to that of the OJ Simpson trial: we aren't interested in knowing "if you have two children, what is the chance they will both die from crib death?" (which is how you get to 1 in 73 million) but "two children died from crib death; what is the chance that they were murdered?" To answer that question properly, you would have to get data on all families who had two crib deaths and see the proportion of families in which the deaths were due to murder.

A key point is that even very unlikely events will happen every so often. One in 73 million is roughly the same chance as tossing 26 heads in a row. Now if you ran into a friend on the street, and he told you that he'd just thrown 26 coins and they'd all came up heads you'd think, "That is very, very unlikely; he's probably lying." On the other hand, if there was a competition to have every adult American throw 26 coins, and it turned out that Joe in Peoria had done it, well that wouldn't be that surprising at all (if you are interested, the chance that at least one out of 300 million Americans would throw 26 consecutive heads is about 98%). And no one would arrest Joe for fraud on the grounds that what he was claiming he had done was very unlikely, and so he probably hadn't. There are lots and lots of families with two children and at least some of them will suffer the tragic and highly unlikely loss of both from crib death.

Act 3: George goes deep

While statisticians don't spend much time worrying about the guilt or innocence of individual defendants in court cases, they do commonly evaluate medical tests. Let's imagine that there was a test for heart disease that had been found to be 99% accurate. Let's also imagine that having run past me into the end zone for a score, George jogged off the Ultimate field and straight into his doctor's office to take the heart disease test. Our statistical question is this: If the test comes back positive, what is the chance that George has heart disease?

The obvious answer is 99%, because the test is 99% accurate. But just like OJ Simpson's lawyers, and Professor Meadows, this argument ignores crucial information. OJ Simpson's lawyers didn't take into account that Nicole Brown was murdered and Professor Meadows ignored the fact that Sally Clark's children were already dead. If we believe that George has a very high chance of heart disease, we are forgetting that he just played two hours of Ultimate and then caught the game's last score.

To work out George's chance of heart disease given that his test was positive, we need to know the typical rate of heart disease in healthy, non-smoking, athletic men in their 20's. A reasonable estimate might be a rate of 1 in 10,000. What happens if we were to give our test to 10,000 men like George? We might first assume that our test does accurately identify the one guy with the heart problem. This leaves 9999 without heart disease. If the test is 99% accurate, it is wrong 1% of the time, meaning that 1% of 9999—about 100 healthy men—will have a positive result. So if we test 10,000 healthy, non-smoking, athletic, 20-something males, we will call 101 positive, but only 1 will actually have heart disease. George's chance of heart disease having been given a positive result by a test that is 99% accurate is about 1%. (This is the *positive predictive value*; see *I ignore my child's cough, my wife panics: About specificity and sensitivity*.) You can show these results in a simple table:

	Has heart disease	Does not have heart disease	Probability of heart disease
Test comes up positive	1	100	~1%
Test comes up negative	0	9899	< 0.01%

Ok, but what if I tell you that the reason George went to the doctor is that he had a history of heart disease in his family and young men with this particular family history have about a 1% chance of heart disease, even if they are athletic? Here is the result you would get for the 99% accurate test:

	Has heart disease	Does not have heart disease	Probability of heart disease
Test comes up positive	99	99	50%
Test comes up negative	1	9801	0.01%

You can see that, out of the men who test positive, half have heart disease. So George's chance of heart disease having being given a positive test result changed dramatically from 1% (which he could basically ignore) to 50% (which means he would really need some extra tests). In other words, the probability of disease *after* you take a test depends on your probability *before* you take the test (the *prior probability*).

OJ, Sally and George meet for coffee and talk about old times

Any statistician hanging around with a latte would be delighted to inform OJ Simpson, Sally Clark and George that what links their stories is the idea of *conditional* probability. You'll hear it said that much of statistics is about giving you the probability of something. For example, a statistician might analyze cancer trends to calculate that a woman's lifetime probability of being diagnosed with breast cancer is about 1 in 8. But statistics is perhaps most particularly concerned with working out a probability given some additional information (the "condition"). The most obvious example is the sensitivity of a medical test which is the probability that a patient comes up positive if they do indeed have the disease (see *I ignore my child's cough, my wife panics: About specificity and sensitivity*). A statistician would describe this as "the probability of a positive test conditional on disease." Conditional probability is generally much more useful than unconditional probability because you are using more information. Bayesian statistics is a branch of statistics that explicitly tries to use as much information as possible, by using prior probabilities to calculate the chance that something is true or false. As a result, Bayesian formulas include conditional probabilities.

A final point: the justice system generally does work out in the end (OJ is in jail and Sally Clark was released). Something that also works out in the end is the aging process, so here is a warning to you, George: you might not have heart disease, but wait 20 years and some kid fresh out of college will burn you in the end zone. And your knee will hurt, and maybe your grasp of math won't be so great, and no, I'm not bitter . . . keep the darn disc.

• Things to Remember •

1. Much of statistics involves working out probabilities.

2. One common error is to calculate the probability of something that has already happened, and then come to conclusions about what caused it based on whether that probability is high or low.

3. The probability of something (such as having heart disease) given information that something else is true (such as the result of a heart test) is called a conditional probability.

4. Conditional probability depends on both the probability before the information was obtained (the *prior probability* of heart disease) and the value of the information (such as the accuracy of the heart test).

Discussion

1. I described two events as independent if information about one gives you no information about the other. Similarly, two variables are independent if information about one gives you no information about the other. What is the relationship between independence and statistical tests such as the *t*-test or chi-squared?

2. In the text I describe Professor Meadows' argument as "only about 1 in 8500 babies die of crib death . . . the chance of two crib deaths [is] 1 in 8500 multiplied by 1 in 8500 [or] 1 in 73 million." I pointed out that the "1 in 73 million" number (a) assumes that the chance that an infant will die of crib death is independent of the chance that a sibling would; (b) says very little about the probability that the children were murdered. There is an even more fundamental mathematical problem though. Any thoughts?

3. The Sally Clark case has been described as an example of the "Prosecutor's Fallacy." What do you think this is?

NOTE: See page 196 for answer sets.

CHAPTER 28

Boy meets girl, girl rejects boy, boy starts multiple testing

Boy: *Would you like to have lunch one day, just you and me?*

Girl: *No.*

Boy: *I thought you said I was cute.*

Girl: *I never said that!*

Boy: *But you laugh at my jokes.*

Girl: *Because they are so stupid.*

Boy: *Is there nothing you like about me?*

Girl: *Well, your fashion sense isn't entirely misplaced.*

Boy: *(hopefully) So you will go on a date with me?*

Girl: *No.*

What is happening here is that the girl clearly doesn't like the boy, but the boy keeps on fishing around, looking for something positive. He eventually finds it, but that doesn't change the fact that his dry spell isn't going to end any time soon.

As such, the conversation has a lot in common with many statistical analyses: the investigators run an experiment (boy meets girl); the results don't come out they way that they hoped (girl rejects boy); the investigators ask the statistician to run more and more analyses to see if they can find something positive (boy learns that girl doesn't mind his clothes).

One of the best examples of multiple testing is subgroup analysis. Let's say we do a clinical trial to see whether a new chemotherapy drug helps prevent recurrence after surgery for colon cancer, with 1000 patients receiving the new drug, and 1000 patients receiving standard therapy. Here are the results we get:

	New drug	Old drug
Patient recurred	150 (15%)	190 (19%)
Patient cancer free	850 (85%)	810 (81%)

The new drug appears to reduce recurrence rates from 19% to 15%. This difference is statistically significant ($p = 0.020$, if you are interested). However, the investigators then think, "Ok, the drug seems to work, great. But we don't just want to tell doctors simply to go use it; they'll ask us who exactly to give it to. Let's see if it works better for some patients than for others."

This seems pretty sensible because medicine should be all about the individual patient, rather than treating everyone the same. So let's imagine that the investigators want to know whether the new drug works better for men or women—that is, they want to know about drug effects in subgroups of patients. Here are their results, separately by gender:

	Men		Women	
	New drug	Old drug	New drug	Old drug
Patient recurred	75 (15%)	100 (20%)	75 (15%)	90 (18%)
Patient cancer free	425 (85%)	400 (80%)	425 (85%)	410 (82%)

The results are pretty similar, apart from a slightly higher recurrence rate in men on the older drug (20% in men, 18% in women). But this slight difference has a big effect on the p-value: there is a statistically significant effect of the drug for the men ($p = 0.046$) but not for the women ($p = 0.2$). The investigators (although not the drug company) are delighted with these results and rush to publish the finding that the drug works for men but not women. However, at the last moment, a research assistant who has been checking the data rushes in with news of a few errors: it turns out that some men had been mislabeled as women and vice versa. The corrected data look like this:

	Men		Women	
	New drug	Old drug	New drug	Old drug
Patient recurred	80 (16%)	100 (20%)	70 (14%)	90 (18%)
Patient cancer free	420 (84%)	400 (80%)	430 (86%)	410 (82%)

When you run these numbers you get $p = 0.12$ for the men and $p = 0.10$ for the women (the overall p-value for all patients is still $p = 0.020$). The investigators conclude that the new drug is effective, but doesn't work either for men or for women. This is impossible, in the same way that $0 + 0 = 1$ is impossible.

How to make an ineffective drug work just fine

It is also impossible that $0 + 1 = 0$. Nonetheless, many medical researchers insist that this is sometimes the case. Take the following study:

	New drug	Old drug
Patient recurred	165 (16.5%)	190 (19%)
Patient cancer free	835 (83.5%)	810 (81%)

The p-value here is 0.16 so we can't conclude that the new drug is any better over our regular treatment. As you can imagine, this is a bit of a crushing blow to the investigators, who have spent years on the study. So they say: "Ok, perhaps the new drug doesn't work on average, but perhaps it works for a subgroup of patients." Here are the results separately for men and women:

	Men		Women	
	New drug	Old drug	New drug	Old drug
Patient recurred	75 (15%)	100 (20%)	90 (18%)	90 (18%)
Patient cancer free	425 (85%)	400 (80%)	410 (82%)	410 (82%)

Now we get $p = 0.046$ for men and $p = 1$ for women. The investigators (and now the drug company too) are delighted and want to say that men, though not women, should be treated by the new drug.

So men are from Mars, women are from Venus?

There are a couple of problems here. First, the investigators appear to be accepting the null hypothesis (see *Michael Jordan won't accept the null hypothesis: How to interpret high* p-*values*). In the first example, they said that the drug didn't work for women because the difference between groups was not statistically significant. But in a drug trial, a p-value of 0.05 or more means "we didn't find sufficient evidence that the drug is different from control," not "the drug is the same as control."

Second, while we might be sympathetic to the investigators' desire to find out whether the drug works better for some patients than others, the particular subgroup analysis they are doing is a little silly. Chemotherapy drugs are basically poisons that are slightly better at killing cancer cells than normal healthy cells (this is why chemotherapy causes unpleasant symptoms such as nausea and hair loss). There is really no reason why a poison would work differently in a man than a woman. I have asked some cancer doctors, and they told me that they couldn't think of any study ever done showing that a chemotherapy agent had a different effect on men compared to women. This was, incidentally, *after* they had just shown me an analysis of the effects of chemotherapy separately by gender.

The third problem, and the one that gets statisticians most excited, is related to the role of chance. Take the first set of chemotherapy study results: the recurrence rates in patients receiving the new chemotherapy drug were the same in men and women; the overall recurrence rate for those on the standard treatment was 19%, but slightly higher for men (20%) and slightly lower for women (18%). These slight differences could quite clearly be due to random variation.

Astrology, and why Geminis aren't helped by aspirin

The problem with multiple testing is that the more significance tests you do, the more likely one will come up significant by chance. If I flip coins every day and look at my results over the course of a year, I probably won't find that I throw more heads than expected by chance. But if I start analyzing my results by time of day, date and weather, it wouldn't be surprising if I found, say, more heads than expected ($p = 0.002$) on wet Wednesday mornings in October. This is exactly the same as the boy asking the girl if she liked him in any way and eventually finding one thing she was okay with, although she still wouldn't go out with him.

In the case where there was no significant difference between the drug group and the control group but the drug did seem to work in men, we did three significance tests (one including all patients, one for men, one for women). If we went really hard at it, and looked at subgroups of gender, age and race, we might end up with 15 significance tests (1 overall, 2 for gender, 2 for age, 2 for race, 8 for combined subgroups, such as older, male, African-Americans). The following table shows the probability that you'll get at least one statistically significant result if you do multiple different tests and the null hypothesis is true (if you are interested in how I worked out this probability, see the discussion section):

Number of tests	Probability that at least one test will be statistically significant at $p < 0.05$
1	5%
2	9.8%
3	14.3%
4	18.5%
5	22.6%
10	40.1%
15	53.7%
20	64.2%
25	72.3%
50	92.3%
100	99.4%

So if we just do one significance test, we have a 5% chance of saying the drug works when it doesn't, which is exactly what we hope for when we chose 5% for statistical significance. If we do three tests, we nearly triple the chance of falsely rejecting the null hypothesis. By the time we get to race, age and gender—15 tests—we have a better than even chance of recommending an ineffective drug to at least somebody (as it happens, these numbers aren't exactly correct for subgroup analysis—see the discussion section—but they are in the right ballpark).

One of the most vocal opponents of subgroup analysis was a British statistician call Richard Peto. He once submitted a clinical trial to a very prestigious medical journal called the *Lancet* showing that aspirin helped after heart attack (this paper is one of the reasons why aspirin is now a standard treatment for heart attack patients). The editors wrote back to say, yes, great study, and we accept that aspirin helps; however, could Peto conduct some analyses to see whether aspirin works better in some patients than in others? Peto said he would do so, but only if he could also conduct some subgroup analyses by astrological sign. As a result, you can look up a paper in one of the best medical journals in the world showing that heart attack patients born under Libra or Gemini don't respond to aspirin.

The problem here is one of statistical power which, just as it sounds, is your ability statistically to do what needs to be done (see *Meeting up with friends: On sample size, precision and statistical power*). Let's get back to our chemotherapy example. Imagine that we wanted to find out if the new drug made any difference to, say, white males aged 40–50, no positive lymph nodes, and a fondness for country music. There are 20 patients in the trial who meet this description, half of whom received the new drug. It is very unlikely that you'd see a statistically significant difference between groups with so few patients. In fact, you'd only get $p < 0.05$ if 5 out of 10 patients recurred on one drug and none recurred on the other.

The following graph shows the probability of a statistically significant result by the number of patients in a trial (or subgroup), if the true difference were a 20% recurrence rate on the old drug and a 15% recurrence rate on the new drug. Because the null hypothesis is false, this probability is the power of the study (see *Meeting up with friends: On sample size, precision and statistical power*). You can see that, even if the drug was effective, there is only a very small probability that the subgroup analysis with the 20 country music lovers would reject the null hypothesis.

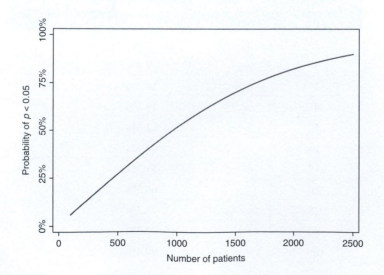

So if you had an uncle with colon cancer who loved country music, was aged 40–50 and so on, I recommend that you ignore the subgroup analysis. Look instead at the overall results of the study and suggest that he use the new chemotherapy drug. If you disagree, then I suppose you'd also be somewhat confused by the lack of astrology charts in the emergency room of your local hospital. Similarly, when considering whether a romantic interest likes you or not, I suggest you look at the big picture and avoid focusing on that one time they may have said something vaguely complimentary.

• Things to Remember •

1. The more statistical tests you conduct, the greater the chance that one will come up statistically significant, even if the null hypothesis is true.

2. A small study has a good chance of failing to reject the null hypothesis, even if it is false.

3. Subgroup analysis involves doing a large number of tests, some of which will involve a small number of individuals.

4. As such, subgroup analysis increases both the risk of falsely rejecting the null hypothesis when it is true and falsely failing to reject the null hypothesis when it is false.

5. If your p-value is greater than 0.05, you don't accept the null hypothesis (e.g., "the drug doesn't work"), you fail to reject the null hypothesis.

6. Take your medicine even if you are born under Gemini or Libra.

SEE ALSO: *Michael Jordan won't accept the null hypothesis: How to interpret high p-values; Some things that have never happened to me: Why you shouldn't compare p-values*

for Discussion

1. Is subgroup analysis a problem mainly for medical research?

2. Do all clinical trials have equal numbers of patients in each group (e.g., 1000 patients on the new drug and 1000 patients in the control group)? Do clinical trials include exactly equal numbers of men and women?

3. *For enthusiastic students only:* One table in this chapter shows the probability that at least one of a given number of tests will be statistically significant if the null hypothesis is true. I worked out these numbers using the formula $1 - 0.95^n$ where n is the number of tests. This is the same as saying, "What is the probability that all tests have p-values ≥ 0.05?" (which is 0.95^n) and then saying, "The chance that at least one test has $p < 0.05$ is 1 minus that probability, i.e., $1-0.95^n$." As I hinted in the text, this formula is an oversimplification. Why?

NOTE: See page 198 for answer sets.

CHAPTER 29

....

Some things that have never happened to me: Why you shouldn't compare *p*-values

Here are some things that have never happened to me:

1. Friends of ours come over for dinner and we are having a nice time—sitting around chatting and laughing, with the kids playing in the next room—when the guy says, "RuthAnne and I would like to compare our *p*-values with yours."

2. A colleague calls me, clearly in a bad way, and wants to go out for coffee. He tells me that he has been having problems at work and, well, one thing led to another, and eventually his boss asked him into her office. She tells him that his *p*-value was one of the largest in his department and that he is at risk of losing his job.

3. I get a flyer from one of the candidates for mayor. The candidate compares his own record on p-values ("Lower p-values for stronger communities") with that of his opponent ("Brown: wrong on p-values, bad for New York").

Normal people might not compare p-values, but scientists do, all the time.

1. Some studies are conducted on whether it takes longer for African American males to hail a taxi than whites. A sociologist reviewing the data argues by stating, "In the Chicago study, Bloggs and colleagues found a clear effect of race on wait times ($p < 0.001$); Smith and colleagues in New York did report longer waits for African Americans, although this effect only just met statistical significance ($p = 0.045$). This suggests that there is a stronger effect of race in Chicago than in New York."

2. From a survey on attitudes to violence in movies, it is stated that "There was a strong association between age and attitude, with older participants more likely to agree that there is too much violence in movies. However, this effect was more pronounced in women ($p = 0.002$) than in men ($p = 0.005$)."

3. An engineer is examining a new heat treatment for engine parts. He concludes by stating, "Overall, the novel treatment significantly decreased failure rate ($p < 0.0001$). However, treatment was more effective for moving ($p = 0.0008$) than for static parts ($p = 0.02$)."

Before we go on, write down some reasons why there might be differences in racial attitudes between New York and Chicago, why attitude to movie violence changes more with age in women than men, and why a new way of treating metal has a bigger impact on moving engine parts than on parts than don't move.

Done that?

In all three examples, the scientist concluded that something had a bigger effect because it had a smaller p-value. To see whether or not this is a good idea, have a look at this data set comparing treatments for high blood pressure:

Drug	Number of patients	Mean decrease in blood pressure	Standard deviation
Standard	10	5	1.5
New Drug A	10	6.5	0.5
New Drug B	10	15	12.5

Which drug would you advise a patient with high blood pressure to take? It seems pretty obvious that you'd recommend drug B—it leads to the largest mean decrease in blood pressure and, in fact, it is significantly superior to drug A ($p = 0.046$). But the p-value comparing each of the new drugs with the standard is actually smaller for new drug A ($p = 0.008$) than for new drug B ($p = 0.022$). If you chose a drug based on comparing the p-values you'd end up with the wrong treatment.

Comparing p-values actually makes no sense at all. First, p-values are about inference, not about estimation. A p-value tells you about the *strength of evidence,* not the *size of the effect*. In the blood pressure study, drug A had stronger evidence even though drug B had a larger effect. Moreover, you need a single p-value to test a hypothesis, not two (see also *Weed control for p-values: A single scientific question should be addressed by a single statistical test).* For the taxi study, the authors drew a conclusion about Chicago and New York that was, in effect, rejecting the null hypothesis that there is no difference between the two cities with respect to the effects of race on hailing a cab. But where was the *p*-value to test this hypothesis?

Which brings us to that list of reasons you wrote down why race-relations are better in New York, why women's attitude to movie violence changes with age more than men's and why heat treatment affects moving engine parts more than parts that don't move. I asked you to write the list to make a simple point: we can come up with explanations for almost anything, but that doesn't make it true. This goes to show that scientists should be more like normal people, and refuse to compare *p*-values.

• Things to Remember •

1. *P*-values measure strength of evidence, not size of an effect.

2. To test a hypothesis (such as which of two new drugs is better) you need a single *p*-value (such as from a *t* test comparing the two drugs), not two separate *p*-values addressing entirely different hypotheses (such as whether each drug is better than a standard treatment).

3. Don't compare *p*-values.

for Discussion

1. Why might a stronger effect lead to a higher *p*-value, and less evidence against a null hypothesis of no effect? How might you explain the differences in *p*-values for the four examples in the text (taxis, movies, engine parts and blood pressure drugs)?

2. How would you test whether one effect was stronger than another? For example, how would you test whether women's attitude to violence in movies changes more with age than men's?

NOTE: See page 200 for answer sets.

How to win the marathon: Avoiding errors when measuring things that happen over time

A friend of mine is an avid runner who trains hard to post a good time in the New York City marathon. Indeed, I remember that we once headed back from the beach early because she had planned a 20 mile run and didn't want to throw off her training schedule. Being a statistician, I know a much easier way to shave off a couple of minutes from a race: just don't start your stopwatch until you've been running for a bit.

Now if that sounds like cheating, just consider the following graph, which shows the results of a study on job satisfaction and employment. Roughly what happened in this study is that new employees at a company were given a questionnaire to assess how they felt about their job. At the end of initial training, which takes place over the course of three months, they were given a second questionnaire. The company management was particularly interested in employees whose opinions improved after their training, whom they called "responders" to training. They hypothesized that responders (gray line) would stay with the company longer than those who were not classified as responders (black line).

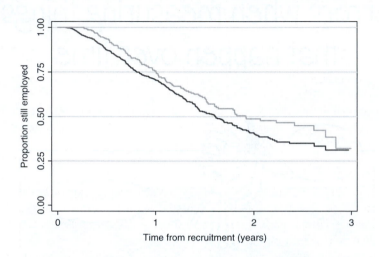

The x-axis gives the number of years since recruitment and the y-axis gives the proportion of employees still working for the company. You can see, for example, that at two years, only about 40% of those who didn't respond to training were employed by the company compared to roughly 50% of responders. If we compare groups statistically using what is known as a log rank test, we get a p-value of 0.026, from which we might well conclude that yes, if you respond to training, you'll stay with the company longer. This might be seen as a justification for the work of the training department.

The problem with this analysis is that you can't "respond to training" if you quit before you complete it, so anyone who quits early is automatically defined as not responding. Indeed, if you look carefully, you'll notice that no responders quit until they have been on the job three months, which is when the follow-up job satisfaction questionnaire is given. Now, I know for a fact that response is not associated with outcome in this data set because I created it artificially using random numbers, altering the data so that any (imaginary) employee who quit before three months was counted as not responding, even if they would have responded had they stayed on. The analysis giving the statistically significant p-value came to the wrong conclusion because the clock was started at a different time for responders than for non-responders.

Two other examples of starting the clock at the wrong time are so common and problematic that they have been given their own special names: *intention-to-treat* and *lead time bias*.

Rock out, it's lead time! (Or: How to live longer but die at the same time)

Rock and roll fans might know that a "lead" is when a guitarist cranks up his amp, plugs in his fuzz box and spends the next couple of minutes rolling around the stage making a lot of noise. As you might guess then, the first time I heard *lead time* during a statistics seminar I got quite excited and started limbering up for some extended facial grimacing. Then followed a discussion about prostate cancer, during which time no one even mentioned Frank Zappa, and my professor kept on saying that lead time was a good thing but had to be avoided at all costs.

When statisticians discuss lead time, they are more likely to be referring to disease screening than Deep Purple. To explain each half of this sentence in turn: Deep Purple are a hard rock band from the early 70's, famed for "Smoke on the Water," a riff that pretty much anyone can play using only the E string of a guitar; disease screening refers to tests that doctors give patients even if they appear healthy. A well-known example is the "Pap" smear, which is a test for cervical cancer recommended to all women, whether or not they have evidence of cancer.

Lead time is the period between when you can catch a disease early using a screening test and diagnosing it later because the patient has symptoms. This sounds like a good idea, because diseases are generally easier to treat if they are found early. This is certainly the case in the disease I study most: an early stage prostate cancer can be cured by surgical removal of the prostate; once the cancer has grown large enough to cause symptoms, it has often spread to other parts of the body and the patient will likely die of his disease.

But a quick cautionary note before you sign up for prostate screening: having your prostate out is no picnic and you can be left incontinent and impotent. A long lead time might well mean "lots of time to catch the prostate cancer early," but it also means "lots of time to die before you realize you have cancer." This means you could go for surgery, miss the opportunity to spend your sunset years enjoying sex and dry underpants, and then die of a heart attack well before the cancer would have caused any problems.

Lead time also causes problems for statistical analysis, particularly for what are called screening studies. A few years back, a group of researchers reported excellent survival rates in patients who had lung cancer discovered by a special type of x-ray, known as a CT scan, and were then treated surgically. The CT scan was given to all smokers, regardless of any symptoms, and was thus a screening test. As the survival rate in lung cancer is generally dismal, the researchers concluded that screening was effective. The problem is that they may well have concluded the same thing had the patients not been treated at all, and screening can't help you if a positive test doesn't lead to an effective treatment.

The following figure shows the progression of cancer from the first few tumor cells to death from disease. Even if no treatment is given, or if treatment is totally ineffective, there is a longer period of time between diagnosis and death if the cancer is screen detected (e.g., found on a CT scan) than clinically detected (e.g., found by a doctor in a clinic trying to work out why a patient is feeling unwell). The clock is starting at a different time for the screen detected cancers than for those detected clinically—this difference is equivalent to the lead time, and biases the comparison between survival rates.

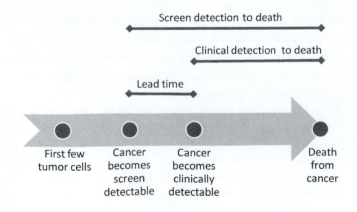

Thus *lead time bias* doesn't mean "the other guitarist is stealing my solo." It means "apparent increases in survival because disease is found early," just like the marathon runner starting the clock at the wrong time.

Getting a treatment even if you didn't:
Intention-to-treat analysis

Let's say you were doing a study to see if counseling reduced the likelihood that a prisoner would end up back in jail after being released. What you might do is select some convicts at random to go to counseling and see what proportion were subsequently jailed in comparison to those not selected for counseling (controls). The problem is that counseling takes some time to set up, and it is quite possible for an ex-con to get in trouble and end up back in jail before you've even worked out which counselor has a free slot.

Typical results of your study might be something like this:

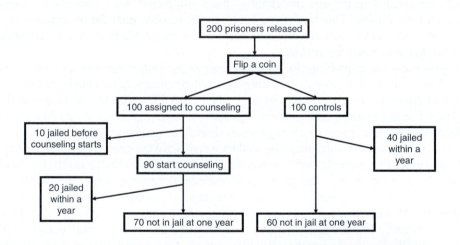

The obvious question is what to do with the 10 who committed crimes and were jailed before their first counseling session. One line of thought would go:

- The 10 individuals who reoffended early never got counseling.
- It is silly to count them in the counseling group.
- It isn't really 30 of 100 in the counseling group who were jailed, it is 20 out of 90 (22%).
- 22% is a lot lower than the 40% who reoffended in the control group ($p = 0.012$), so counseling clearly makes a difference.

The problem with this argument is that convicts who are jailed in the control group are included in the analysis no matter when they get caught breaking the law. This is none other than our marathon runner cheating: we are starting the clock at a different time for the control group (immediately on release from jail) and the counseling group (only once counseling has started). Statisticians say that the fairest analysis is *intention-to-treat*: it doesn't matter what actually happened to you, it was what was intended that counts. In an intention-to-treat analysis you compare a 40% reoffending rate with 30%. In this particular case, you get a smaller difference that is not statistically significant.

Intention-to-treat makes sense, because we can't decide what actually happens, we can only decide our intentions. A beach trip is a simple example. Let's say that I am sitting at home with my family and we are wondering whether to go to the beach. Part of the issue is that we sometimes make a huge effort to get out of the house (round up towels and swimming stuff, put on sunscreen, make snacks, etc.) and then load the car but never get to the beach because traffic gets very bad or because it starts raining. Imagine that we had a data set where we worked out what we'd done and rated each day:

What happened?	How was the day?
Stayed at home	Good
Went to beach	Good
Tried to go to beach, sat in traffic for an hour, then it starting raining so we turned around and went home	Bad
Stayed at home	Bad
Went to beach	Good

On days where we got to the beach, we had a good day 100% of the time compared to 33% of days when we didn't get to the beach. But our decision isn't whether to be at the beach or not, our decision is whether or not to *try* to get to the beach. In order to inform our decision ("Should we try to go to the beach?"), we really have to count the day where we tried and failed to get to the beach as a beach day: this gives 67% of days we tried to go to the beach being good ones compared to 50% of days when we didn't try to go to the beach.

So you count a day as a "beach day" even if you didn't get to the beach and a convict as having received counseling even if he didn't. This sounds all a bit paradoxical, until you realize that the alternative is not much different from the marathon runner starting her stopwatch halfway through the race.

• **Things to Remember** •

1. Many things we analyze statistically take place over time, whether employees leaving a company, cancer growing in a patient, or even just a trip to the beach.

2. A statistical analysis of something that changes over time involves a decision of when to "start the clock."

3. Many statistical errors occur because of starting the clock at the wrong time.

4. A common analysis is to assess the association between a predictor (such as a job satisfaction questionnaire) and time to something happening (such as quitting a job). A typical mistake for such an analysis is to start the clock before you would have information on the predictor (such as measuring time from recruitment, when employees don't complete the questionnaire until they have been on staff for three months).

5. Often you have some kind of problem (such as cancer) that leads to an end result you want to avoid (such as death). If you find a way to find the problem earlier, then the time between the problem and the end result will inevitably be longer. This is known as *lead time bias* and doesn't mean that finding the problem early did any good.

6. Sometimes you try to do something (counsel prisoners, go to the beach) to achieve a particular goal (avoid prison, have a nice day). To work out whether this is worth it, you have to look at all times when you tried (e.g., packed the car) not just those times you succeeded (e.g., got to the beach). This is called *intention-to-treat* analysis.

7. This article is for teaching about statistics. Please don't use this article for information about cancer screening, trips to the beach, the merits of Deep Purple, or artistic tensions in my band.

for Discussion

1. How do you think you should analyze the data from the job satisfaction study?

2. How do you deal with lead time bias?

NOTE: See page 203 for answer sets.

The difference between bad statistics and a bacon sandwich: Are there "rules" in statistics?

I once heard someone describe a statistical analysis as "not strictly kosher." This might well be true, on the grounds that the statistician involved had probably not been certified by the appropriate rabbinic authorities. Nonetheless, leaving aside the issue of whether the Mann-Whitney test counts as milk or meat, the idea of "kosher statistics" does give a wonderful insight into how many scientists view statistics: a set of laws, handed down from above, violation of which constitutes a transgression. As a statistician, I am repeatedly asked whether a particular statistical analysis is "allowed" or whether it would be "against the rules"; as a statistics teacher, my students' questions often concern "Right and Wrong."

It is hard to think of any other area of science that is characterized with so many religious and legal metaphors. We don't wonder whether, say, use of an inappropriate questionnaire would break any laws or whether failure to clean laboratory equipment thoroughly is an eternal or just a venal sin. In short, the way that many scientists understand statistics is deeply unscientific.

One of science's defining characteristics is that new ideas are developed and that both new and old ideas are tested empirically. This is as true for statistics as for any other science. Many of the techniques I use in my day-to-day work—Cox regression, bootstrapping, k-fold cross validation, general estimating equations—were invented relatively recently (I won't explain these any further, as they are beyond the scope of an introductory textbook). I myself have developed a new statistical technique, decision curve analysis, and I'd be happy to explain this to any reader experiencing insomnia.

Moreover, statisticians test methods experimentally: we have computers simulate data sets, then apply different statistical methods and see which come up with the right answers. If I recall correctly, the scientist who liked his statistics kosher was concerned about the use of a t test on skewed data (see *A skewed shot, a biased referee*). Statisticians have tried applying the t test to data simulated so that they are skewed and have found out that, in some cases, it results in p-values that are too high. The study that this scientist criticized reported a statistically significant difference between groups. Accordingly, it doesn't matter that this p-value was likely too high, because the null hypothesis was rejected.

So, no, the t test was not inscribed on the stone tablets Moses brought down from Mount Sinai along with the commandment "Thou shalt not use with skewed data." The t test was invented by a statistician (a guy who worked for a beer company) and has subsequently been tested by other statisticians (including me) to find out if it is any good (it turns out it isn't ideal for the applications I need and I rarely use it). Just like any other science, what you want to know about any statistical technique is the degree to which it might give you a wrong answer and whether there are other methods around that give you a better chance of getting things right. There aren't rules, laws and commandments about this, you just have to know the latest research data.

• Things to Remember •

1. Statistics is used to help scientists analyze data, but is itself a science.
2. Just like any other science, statistics isn't subject to a set of unchanging rules.
3. Statisticians choose methods because of evidence that the methods are helpful for getting the right answer
4. Statistical methods aren't specified in some holy book, they are developed and tested by statisticians.
5. I am Jewish, but I have to admit that I have not had my statistical software certified as *glatt* by a rabbi.

for Discussion

1. So there is no right and wrong in statistics? Does that mean that anything goes and you'll get 100% on your exam even if you do a dumb analysis?

2. Many textbooks (as well as a good number of statistics teachers) say that you should avoid the *t* test for skewed data. Are they wrong?

3. In the text I said that there aren't rules, laws and commandments about statistics, you just have to know what the latest statistical research shows. Does this suggest that all scientists need to look up the statistical journals before running a simple statistical analysis?

NOTE: See page 204 for answer sets.

Look at your garbage bin: It may be the only thing you need to know about statistics

Science, statistics and reproducibility

A key characteristic of science is *reproducibility*. Consider: two literature PhD students writing theses about, say, Hamlet, would not be expected to reach similar conclusions, indeed, there would likely be somewhat of a problem if they did. On the other hand, if two molecular biologists wrote dissertations with differing findings about the same cell pathway, their hopes for a tenure-track position would likely go the way of Rosencrantz and Guildenstern.

The reason why the two biologists should be expected to come to similar conclusions about a phenomenon is related to a second key characteristic of science, *the systematic attempt to avoid error*. We wash test tubes, calibrate instruments and tape-record interviews because we have found out that, if we don't, we often get misleading results.

Here is a quick guide to statistical procedures as implemented by a typical non-statistician.

1. Download the data set from the research database into a spreadsheet.

2. Notice a couple of errors in the data set, make corrections directly onto the spreadsheet.

3. Cut and paste in a couple of columns of data from a different spreadsheet.

4. Notice some missing data, pull research notes and type what is missing onto the spreadsheet.

5. Use spreadsheet functions to create some new data columns, for example, create an "age" variable by subtracting "date of birth" from "date started study."

6. Cut and paste the spreadsheet into a simple software package.

7. Delete (or lose) the original spreadsheets.

8. Use the pull-down menus on the statistical software to run some analyses.

9. Cut and paste the analyses into a word processing document for the journal paper.

It is not hard to see how error could be introduced at just about every stage of this process—indeed, the analysis could probably not be reproduced at all. This leads to what I'll call *J-Com's Law* in honor of the worst typing mistake in history: in 2005, a Japanese trader trying to sell 1 share of J-Com stock at 610,000 yen, instead sold 610,000 shares at 1 yen, losing his firm about $225 million in the process. J-Com's law is:

> *Many of the research papers you read will be wrong not as a result of scientific flaws, poor design or inappropriate statistics, but because of typing errors.*

Here are some real-life examples.

- I looked up a paper in a very prestigious medical journal for some research I was doing. The authors reported that the mean change in pain was 16 points, with a standard deviation of 8 points. However, this relatively small standard deviation didn't make any sense in terms of some other numbers given in the paper, such as the *p*-values. When I contacted the authors, they apologized, and said that they meant to give a figure of 18 rather than 8 for standard deviation.

- I read a study reporting that counseling improved anxiety, depression and fatigue in cancer patients. On the other hand, overall quality of life seemed to be *lower* in patients receiving counseling. I wrote to the authors asking how a treatment could relieve a bunch of distressing symptoms, but make patients feel worse. It turned out that quality of life was indeed better after treatment but that a minus sign had been omitted from the table of results.

- I was sent a data set from colleagues at another institution. When I started my analysis, I immediately noticed some anomalies. The results included data on two proteins which are inversely correlated (that is, you normally see high levels of one or the other but not both) yet some patients had high levels of both proteins. I asked the investigators about this, and they said that they had checked the records for these patients, had found the data to be correct, and that "double-positives" sometimes happened. This seemed reasonable, until I looked at some other variables which were normally positively correlated with one of the proteins. There were several patients who had some odd looking data, and these turned out to be the same "double-positive" patients we had asked about previously. When I raised the issue again, the investigators wrote back saying that, on further checking, there had been data entry errors and that the values of the proteins had been switched in some cases.

- I was asked to help a surgeon conduct an analysis of the effects of obesity on complication rates. The very first line of the spreadsheet he sent to me described a female patient who was 6 feet tall and weighed 130 pounds, roughly the look favored by the typical fashion magazine. Yet her body mass index was given as 49, which put her in the category of the super-obese. It turned out that the surgeon had typed numbers from the surgical charts into a web-based body mass index calculator and then cut and pasted the results back into the spreadsheet. Inevitably, mistakes had occurred.

Avoiding error in statistics

If you take a statistics course, you'll probably learn how to calculate a mean, or work out a p-value from an ANOVA; you'll be given formulas, and advice on how to interpret your results. But what is perhaps just as important in coming to a correct scientific conclusion is making sure that you avoid data and typing errors. If science involves "a systematic attempt to avoid error," then conducting a scientific statistical analysis involves setting up some kind of error-catching system.

1. **Avoiding data collection errors**

   a) *Write a set of procedures for checking data as it is collected.* In a questionnaire study, for example, I had research assistants examine each questionnaire as it was received to check that every question had been answered and that any written text was legible and made sense. If there were any problems with the questionnaire, the research assistants were asked to contact the study participant and clarify things.

   b) *Use "sign offs."* We asked research assistants to sign and date questionnaires that they had checked before filing them. The signature serves the same function as the courtroom oath to tell the whole truth: "I have checked this questionnaire and found it complete and correct." It also means that, if a mistake has been made, we could work out who made it—a powerful incentive to research assistants to be as careful as possible.

2. **Avoiding data entry errors**

   a) *The best way of preventing data entry errors is to avoid data entry altogether.* In the case of the surgeon interested in obesity and complication rates, there was no need to do any cutting and pasting: heights and weights could have been downloaded into a spreadsheet or statistical software package directly from the surgery database, and body mass index then calculated using a formula. Questionnaires can often now be optically scanned or put up on the web. Data can then be directly transferred from the scanner or the web to the study database.

   b) *Use double data entry.* Sometimes you have to use paper questionnaires or forms, and have no choice but to type the results into a computer. The best system to avoid data-entry errors is what is known as *double data entry*. In brief, data from paper forms is typed onto a database; a blank copy of the database is then made and the data re-entered;

the two databases are then merged to discover inconsistencies. The last time I did this, there was at least one inconsistency on 14 of the 99 records in the data set.

c) *Write a protocol for data entry.* A protocol specifies rules for data entry such as how to handle illegible or ambiguous data. As a typical example, we might specify that, if someone circles two responses to a question (e.g., "agree" and "strongly agree"), we take the response corresponding to the higher score (i.e., "strongly agree").

3. **Avoiding errors during data analysis**

a) *Create a log file.* Record with dates all the analyses you do, along with their rationale. The log file should also document the names of files and folders you set up to manage your data. When you update a file, create a new copy and label it with a version number (e.g. "Results of Student Survey v3"). Make sure to keep the old version of the file (e.g. move "Results of Student Survey v2" to a folder called "Previous versions of results section").

b) *Check the final data set.* Once you have the data set in your statistical software, you should check for missing data. You should also conduct consistency and range checks. A *consistency* check determines whether the value of one variable is unlikely or impossible given the value of a different variable. As an example, if you found someone in the data set who was a 16-year-old registered Republican, you'd know there was a problem because you can't register to vote before you are 18. Similarly, someone who gave their job as "statistician" and had a high score on an extroversion scale should probably be investigated further. A *range* check determines whether the values of any variable are unlikely. For example, you would probably want to check data suggesting that someone was 161 years old, had a GPA of 38 or took nearly 2 days to run a marathon.

c) *Program your analyses.* An introduction to statistical programming is really beyond an introductory textbook. However, the key point is that, while it might be fine to use pull-down menus for class assignments, any analysis you run for your own research really should be conducted by writing programming code (an important reason to take more statistics classes). One obvious point is that programming helps ensure reproducibility of an analysis—you just run your code again. The code ideally should include automatic output suitable for importing into a word processor—cutting and pasting individual numbers from software output is an important source of error.

4. **Manuscript preparation**

a) *Check every number on your report against the printout from the statistics software.* This offers an additional way of ensuring that the paper says what it is meant to say.

b) *Double-check final copy.* Errors often creep in when papers are reformatted, say, by an editor of a scientific journal.

All this checking can be very time consuming and certainly sounds somewhat depressing that you need to learn statistical programming code if you want to run high-quality analyses. But no laboratory scientist uses dirty equipment on the grounds that bottle washing takes too long. Moreover, in my experience, systematic data checking generally saves time because it prevents problems that are extremely difficult to remedy once they have occurred. Also, I challenge any researcher to say, "Sorry, I was too busy to check the data." On which point, by the principle of "garbage in, garbage out," look at your garbage bin. It may be the only thing you need to know about statistics.

• Things to Remember •

1. Two key characteristics of science are reproducibility and the systematic attempt to avoid error.

2. For a statistical analysis to be scientific, it needs to be reproducible. Writing programming code for an analysis helps reproducibility.

3. Many errors in scientific reports result not from flawed study design, poor experimental technique or inappropriate statistics, but from simple errors in data collection or typing.

4. Scientific statistical analysis needs to include specific steps designed to reduce these errors.

5. In my next book I'll discuss reproducibility in science, along with the implications for statistics.

for Discussion

1. *For enthusiastic students only:* Is there anything I need to know about statistical programming, other than the fact that I shouldn't bring this up on a first date?

NOTE: See page 205 for answer sets.

CHAPTER 33

Numbers that mean something: Linking math and science

It was just before a 7 am meeting and I was really trying to get to the bagels, but I couldn't help overhear a conversation between one of my statistical colleagues and a surgeon.

. . . .

Statistician: *Oh, so you have already calculated the p-value?*

Surgeon: *Yes, I used multinomial logistic regression.*

Statistician: *Really? How did you come up with that?*

Surgeon: *I tried each analysis on the statistical software drop-down menus, and that was the one that gave the smallest p-value.*

. . . .

You've got to give the guy some points for honesty. I don't think I've ever heard anyone describe so clearly what is perhaps the most typical approach to statistics:

- Load up the data into the statistics software.
- Press a few buttons.
- Cut and paste the results in a word processing document.
- Look at the p-value: if p is less than 0.05, that is a good thing. If $p \geq 0.05$, your study was a failure and probably isn't worth sending to a scientific journal.

A few years back, I went to a conference on prostate cancer. In one presentation, which was pretty representative of the conference as a whole, a surgeon had loaded data on a group of cancer patients into a basic statistical software package and then selected the appropriate commands to see what was associated with whether the cancer returned ("recurred") after surgery. The surgeon then read down a list of variables and concluded that each "predicted recurrence" ($p < 0.05$) or "did not predict recurrence" ($p \geq 0.05$). One of the variables was obesity, and because the p-value was something like 0.02, the presenter concluded that "obesity may have some effect on survival." (I flew all the way to Atlanta to learn that?)

The analysis that the surgeon conducted is known as a multivariable Cox proportional hazards model. Calculating a Cox model involves some very complicated mathematics and is impractical without a computer. This is exactly the problem.

In my favorite picture of R A Fisher, one of the founders of modern statistics, he is seated at a desk operating a mechanical counting device. Conducting a complex statistical analysis on such a machine is extremely time consuming. Anyone who, like Fisher, had to depend on mechanical calculators would have had to think extremely hard about the analysis they wanted to conduct before they started. With modern computing it is possible to conduct an analysis with a minimum of time (or brain power—you just select something from a drop-down menu). The inevitable result is a proliferation of analyses that have not been sufficiently thought through, leading to uninteresting conclusions such as "we have evidence against the hypothesis that obesity does not influence the course of cancer."

How I would have approached the problem is as follows. Having a high body mass index cannot possibly, in and of itself, affect cancer. (It is not as if the cancer cells think, "Wow, this guy is pudgy, let's go crazy.") There has to be some *reason* why obesity is associated with higher recurrence rates. The first step in any analysis is to work out what these reasons might be. I came up with the following list:

Biology: there is something about the biology of obese individuals that promotes the growth of prostate cancer. For example, it is known that prostate cancer is affected by hormones such as testosterone and insulin, and that levels of these hormones are affected by obesity.

Behavior: obese individuals engage in behaviors that increase the risk of prostate cancer recurrence. It might be, for instance, that obese men exercise less than the non-obese, and that exercise helps suppress cancer; alternatively, fat and sugar in the diets of obese people might promote cancer growth.

Surgery: surgery in obese individuals is more difficult. As a result, surgeons may not remove all cancer tissue; the cells that are left behind might grow and cause a cancer recurrence.

Something else altogether: there isn't a true association between obesity and cancer, it is an artifact of the scientific and statistical methods used.

It is this last explanation that I'd test first. Don't worry too much about the details that follow—there is probably more about prostate cancer here than you would ever need to know. Just keep hold of the general process that I am going through, which is thinking of what might be happening in a patient's body and trying to convert that into a mathematical hypothesis. In the surgeon's analysis, he used multivariable regression to control for cancer severity. In other words, the question wasn't so much "Is obesity associated with cancer recurrence?" but "If you had two men, one obese and one non-obese, and they had prostate cancers that were identical in all respects we could measure, does the obese man have a higher chance of recurrence?" One of the ways we estimate "cancer severity" is to measure a protein called PSA. The higher your PSA, the worse your cancer is thought to be. PSA is measured in nanograms per milliliter, that is, a weight divided by a volume. It seems reasonable that two patients with similar tumors, one obese, one non-obese, will have a similar weight of PSA, but this will be distributed in a greater volume of body fluid in the obese patient. As a result, the PSA value in "weight per volume" will be lower in the obese man. So if our two patients have similar levels of PSA, it is reasonable to suppose that the obese patient has a larger tumor.

All of which is to say that one theory of why obesity is associated with prostate cancer recurrence is that we underestimate the severity of prostate cancer in obese men. This scientific theory then needs to be turned into a statistical hypothesis. In brief, what I would do is multiply PSA by weight to get a rough estimate of the total amount of PSA. I'd then use this as a covariate in the multivariable model in place of the usual PSA measurement. As it happens, a couple of years after I went to Atlanta, some researchers followed a similar approach and found evidence in support of the hypothesis that the apparent effects of obesity in prostate cancer were related to greater "dilution" of PSA in large men.

My approach here was to think about biology, turn it into math and then think how to apply the results of the math back to biology again. This illustrates a key principle of statistics: linking math and science. I am a biostatistician so I link math and biology; an economist links math to economic behavior; statistics in psychology is about linking math to the human psyche.

It is, of course, easier just to shove everything into the statistical software and interpret the resulting p-values as "yes" and "no." And if this is how you want to approach statistics, you'll have plenty of company. But, please, keep it to yourself and don't block the bagels.

• **Things to Remember** •

1. Computers now make it extremely easy to conduct even the most complex statistical analyses at the touch of a button.

2. The printout from statistical software is just a bunch of numbers, many of which won't mean much.

3. What statistics should be about is linking math to science:

   a. Think through the science and develop statistical hypotheses in the light of specific scientific questions.

   b. Interpret the results of the analyses in terms of their implications for those questions.

for Discussion

1. Jonas goes to the market and buys a 10 lb watermelon and 9 apples weighing $2\frac{1}{2}$ lbs. He calculates the mean weight of the fruit as $12\frac{1}{2} \div 10 = 1\frac{1}{4}$ lbs. What are your thoughts about this statistic?

2. The surgeon concluded that "obesity may have some effect on survival." Words like "may," "might" and "could" are often found in the conclusion of scientific studies. Why should scientists avoid using these words?

NOTE: See page 207 for answer sets.

CHAPTER 34

Statistics is about people, even if you can't see the tears

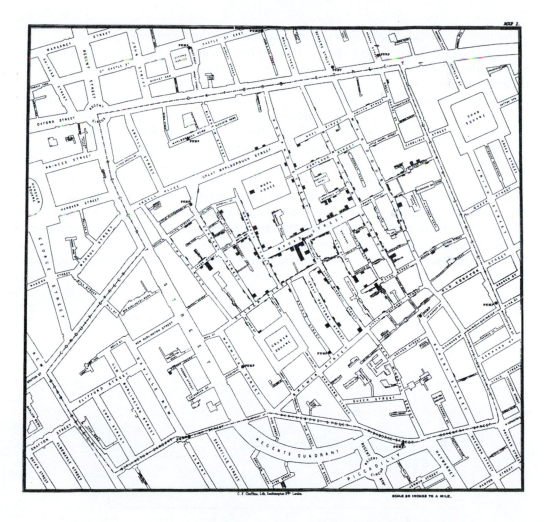

You can read a lot of things about statistics in this book. Chances are you will forget quite a few of them. But please, don't forget this: statistics is about people, even if you can't see the tears.

What you start with when you do statistics is a list of numbers and other data that mean very little on their own. Like this for example:

Address	Number of cases
32 Broad Street	2
34 Broad Street	4
21 Warwick Street	0
23 Warwick Street	etc.

But then with statistics, you can transform those numbers into something meaningful.

The picture on the previous page is a map of cholera cases in mid-19th century London that was plotted by John Snow, a doctor. The map shows that the cases were clustered around the Broad Street water pump. The pump handle was removed and the number of cholera cases declined. It is because of John Snow that cholera no longer sweeps through our cities, killing thousands. If the New York cholera epidemic of 1832 happened today, and killed the same proportion of New Yorkers as it did in 1832, over 100,000 would die. Snow's work was pivotal in recognizing germs as the cause of disease, leading to proper sanitation and clean water, as well as antibiotics and vaccination.

It is all too easy to forget that statistics is really about people. I once asked a well known cancer researcher for some information about a study he had published. He refused, on the grounds that I might use the information to "cast doubt" on his findings. This is clearly someone who cares more about avoiding a minor career embarrassment than about understanding cancer data in order to help future generations of patients.

This cancer researcher had published several scientific papers that included graphs similar to this one:

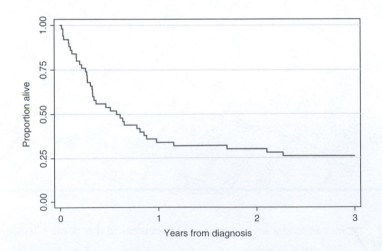

Known as a "survival curve," the graph shows on the y axis the proportion of cancer patients still alive after the period of time given on the x axis. So each little step down on the graph is when a person died. That person is somebody's son or daughter, perhaps also a husband or wife or mother or father. Deaths from cancer are often painful and rarely sudden, which means that each little step on the graph is someone who knew that they were going to die and likely suffered as they did so.

Statistical analysis of this type of data often involves examination of the "proportional hazards" assumption or consideration of "accelerated failure time models," with results expressed in terms of "hazard ratios." Indeed, you could read statistical analyses of cancer data all day and never realize that any real people were involved. So you have to remind yourself of that, always. I try to: when I hear someone say something like: "she's not a statistic, she's someone's little child," I think to myself, "what's the difference? A statistic is *always* someone's child."

I once gave a lecture in a small college town in the Midwest. While I was there, someone asked me what it was like to live in a dangerous big city like New York. It turns out that the police precinct in which I live has almost exactly the same population as the college town. When I looked up the crime figures, it turns out that they have four times as many crimes. (And that isn't even counting the omelette the hotel made me for breakfast.) I could live my life on the basis of New York's reputation. But I choose to live by the statistics instead, which means that I get to raise my family in beautiful brownstone Brooklyn, eat great food in local restaurants, and feel relatively safe as I do so.

Sometimes, when I am walking around in Brooklyn, I look at my children and I find myself thinking about John Snow. Two hundred years ago, before the cholera map, about one in five children would not live to see their 10th birthday. It is now about 1 in 100, a 20-fold difference. Statistics can be hard to learn, and even harder to master. But, as I said right back when we started, we want to live our lives better. To do that, we have to make good decisions, and sometimes looking at numerical data in the right way can help us to do so.

Discussion Section

CHAPTER 1
I tell a friend that my job is more fun than you'd think: What is statistics?

1. **I defined a hypothesis as "a statement about the world that could be tested to see whether it is true or false." Are there some statements that can't be tested?**

   Short answer: yes, of course, many statements can't be tested at all. Which is a good thing, because I hate the idea of having to do statistics for "I love my wife" or "Mozart wrote great operas." That is pretty much all you need to know, so feel free to jump to discussion point 2. If you are feeling philosophical, however, read on.

   A common reason why many statements can't be shown to be true or false is that they are too vague. A great example of this is horoscopes: Capricorns might be told, for example, to be careful in family relationships because it is possible that a close family member is keeping quiet about some personal troubles. It is difficult to think of how this statement could be anything but true because pretty much anything is "possible." Words like "possible" or "may" or "might" are known as "weasel words" because they allow you to "weasel out" of any claim you make (e.g., you say "we shouldn't go to the game because it may rain," so everyone stays home and gets bored. When it doesn't rain you defend yourself saying, "I didn't say it would rain, only that it might."). Statements can also be too vague if it is possible that they are partly true and partly false. This would apply, for example, to statements like "Chiropractic is an effective form of medical treatment" because chiropractic probably works for some problems (such as back pain) but not others (such as the black plague).

   Another reason why a statement might not be able to be disproved is if it is defined so broadly as to be true in all circumstances. Some self-help "New Age" gurus believe that men need to get back in touch with their inner manhood (e.g., by running around in the woods naked under a full moon). A friend once showed me a book promoting this vision of the modern male. Opening it at random, I read something like "The fact is, women like men who fight, and love men who fight well." When I pointed out that this was obviously false—on the grounds that very few men fight at all, and that most men seem to find love, even the wimpy, nerdy types—my friend argued that I was defining "fight" too narrowly, and that "fight" also means standing up for what you believe in, for example, by defending your point of view in a debate. This covers just about everybody.

   The third reason why some statements can't be shown to be true or false is that we may have no methods available to do so. When a psychotherapist states that alcoholism often results from unresolved subconscious conflicts (perhaps an overactive *id* with an underactive *super ego*), we have no way of knowing whether this is true or not because we don't have a tool to measure subconscious activity.

   Some philosophers have argued that what defines science is that a scientific statement can, in theory, be shown to be false (what is known as "falsification"). If a statement can't possibly be shown to be false, then there is no point trying to get evidence one way or the other and so science doesn't enter into it. It turns out that this view can be hard to defend philosophically, but it is undoubtedly a good rule of thumb and provides a quick life tip: if you are arguing with someone who says things that can't possibly be shown to be false, stop arguing and go for a walk or something.

2. **There are two sorts of estimates that statisticians make: how big or small something is and how big or small something is compared to something else. An example of the first sort of estimate is "the mean height of an American male is close to 5 ft $9\frac{1}{2}$ in." An example of the second sort of estimate is "men who smoke are 23 times more likely to develop lung cancer than men who have never smoked." Write down some examples of estimates of both sorts.**

Here are some examples I came up with: About 400,000 Americans are diagnosed with heart failure each year; Dutch men are on average three-and-a-half inches taller than Ameri-can men; a 65-year-old man who has smoked a pack a day since his teens has a 10% chance of being diagnosed with lung cancer by age 75; women earn about 20% less than men doing comparable jobs.

3. **Most hypotheses can be rephrased in terms of estimates. I mixed up some estimates and hypotheses below. Match each estimate with the corresponding hypothesis and say which is the estimate and which the hypothesis.**

   a. Hypothesis: Men and women do not differ in their voting behavior for presidential candidates. Estimate: The proportion of women voting for Democratic presidential candidates is 5% higher than men.

   b. Hypothesis: Chemotherapy plus surgery is no more effective than surgery for breast cancer. Estimate: Recurrence rates were 5% lower in women receiving chemotherapy after surgery compared to women receiving surgery alone.

   c. Hypothesis: Improvements in street lighting decrease crime. Estimate: Crimes decreased 21% comparing the year before and the year after completion of a program to improve street lighting.

   d. Hypothesis: Electric shocks (punishment) are more effective than sugar (reward) for improving learning in rats, as measured by time to complete a maze learning task. Estimate: Mean time to complete a maze was 20 seconds shorter in rats exposed to shocks than those given sugar.

   e. Hypothesis: Obesity rates in California increased during the 1990's. Estimate: Obesity rates in California almost doubled between 1990 and 2000, from 10% to 20%.

One thing you'll notice is that hypotheses are simple statements of fact that can be true or false; estimates have numbers describing how large or small something is.

4. **Who said "there are lies, damned lies and statistics"?**

Just about everyone I have ever met in my professional career. And they all think that it is the first time I have heard the quote, and that they are being really so terribly witty. Also, I don't think that anyone who has quoted me the line has understood it, anyway.

"There are lies, damned lies and statistics" was popularized by Mark Twain, but was first used by Disraeli, a British politician from the 19th century. The point he was trying to make is along the lines of "bad, worse, worst". In other words, statistics are the worst lies you can imagine. What he was referring to was the tendency of politicians to throw around statistics that don't mean much, but which advance their political arguments. A classic modern example is the claim that a presidential candidate had "voted to raise taxes 94 times." To get this number, the candidate's opponent included votes against tax cuts, votes for measures that raised taxes for some and lowered it for others, and votes to close tax loopholes. So one can imagine how Disraeli might describe the "94 tax increases" statistic as being worse than a terrible lie.

The nice thing about "lies, damned lies and statistics" is that it illustrates something very important about the word "statistics." People who smirkingly tell me about "lies, damned lies and statistics" think that it is a smear on my profession, just as would likely irritate sociologists. But "statistics" has two meanings. You can talk about statistics as a science, as in "statistics is used to analyze the results of medical studies." Alternatively, "statistics" is the plural of "statistic," a number obtained from data by statistical analysis. (as in, "We calculated several statistics including the median age and height and the proportion who were women.") When Disraeli complained about "lies, damned lies and statistics," he was referring to this second meaning. As such, he would likely have welcomed better use of the science of statistics to improve the value of political discussion.

CHAPTER 2
So Bill Gates walks into a diner: On means and medians

1. **I said that "half of the sample have values higher than the median and half have values lower than the median." Is that always true?**

    Not always, although it is a pretty good rule of thumb. The rare exception is when you have many observations that have the same value ("ties"). For example, imagine if the salaries in the diner were $30,000; $35,000; $40,000; $40,000; $40,000; $40,000; $80,000. The median salary here is obviously $40,000, leaving only 1 person (14% of the sample) with a salary higher than the median and 2 (29% of the sample) with salaries lower than the median. As another example: most patients go home two days after surgery, although a few stay three days and a small minority, less than 10%, have complications leaving them in hospital for a week or more. The median hospital stay works out at two days, with virtually no one having a hospital stay less than the median and only 20 or 30% with hospital stays longer than the median.

    However, if you have a reasonable number of observations and not too many ties (which is most of the time) then "half of the sample have values higher than the median and half have values lower than the median" works just fine.

2. **Here is a die rolling game: you roll a die and if you get 1–5, I give you $20; if you roll a six, you give me $1000. Would you play? Explain your answer.**

    It is pretty obvious that you should refuse my bet, but it is worth thinking through why, exactly. Let's imagine that we played the game 600 times, wrote down the result of each game and then examined our data set. We'd expect that you would win $20 about 500 times and that I'd win $1000 about 100 times. Your median winnings for such a data set would be $20; the mean would be that you'd lose around $150 ((500 × $20) – (100 × – $1000) = $90,000;

$90,000 \div 600 = \$150$). This illustrates why the mean is generally better than the median for decision making.

CHAPTER 3
Bill Gates goes back to the diner: Standard deviation and interquartile range

1. **The upper and lower quartile are sometimes described as the 75th and 25th centile (or percentile). Explain this.**

   Centiles (or percentiles) are calculated in a similar way to medians and quartiles. To find the 38th centile, for example, you arrange observations from lowest to highest and find a number that is 38% of the way along. As such, 38% of the observations will be less than the 38th centile. Similarly, 62% of the observations are less than the 62nd centile; 92% of observations less than the 92nd centile and so on. We know that a quarter (i.e., 25%) of the observations are below the lower quartile, so the lower quartile is the 25th centile. A quarter of observations are higher than the upper quartile, so three-quarters or 75% are lower than the upper quartile, and therefore the upper quartile is the 75th centile.

2. **I said that, using the mean and standard deviation, you could calculate that "5% had salaries above $61,592 or below $23,128." I could have said: "5% had salaries of $61,592 *or more* or $23,128 *or less.*" Does it make a difference?**

   Means and standard deviations are numbers calculated from the data set using formulas. It is not unusual that no one does (or even could) have a value at the mean, or whatever numbers you calculate as being above or below a certain percentage of the observations. For example, I can calculate from the data set that 80% of the salaries should be less than $50,456.611, which is an impossible salary. If no one has a salary of $50,456.611, then it obviously doesn't make any difference whether you say "80% have salaries less than $50,456.611" or "80% have salaries of $50,456.611 or less."

   If you want to get more technical about it (and if you don't, please skip to discussion point 3), using the mean and standard deviation to calculate what proportion of a data set has values above or below a particular number involves a formula, derived from calculus, based on calculating the area under a normal distribution. Because there is zero area under a point on the curve, it makes no difference whether you say "greater than x" or "at a value of x or higher," so statisticians normally just say "greater than x" to be brief.

3. **Is it really true that 95% of observations are within two standard deviations of the mean, even for a perfectly normal distribution?**

   "Two standard deviations" is just an approximation. The number you should use is really 1.96 or, if you want to get really persnickety, 1.959964. These numbers are calculated by integrating the normal distribution. If you haven't taken calculus or don't feel like working out the integral of the normal distribution (which is a very complicated formula), then don't worry—someone else has already done the calculations for you. You used to be able to buy books of statistical tables that you could use to look up the right numbers. Nowadays, all the numbers are stored as part of statistical software.

CHAPTER 4
A skewed shot, a biased referee

1. How would you avoid selection bias in the surgery study?

The usual way to avoid bias in medical studies is to conduct what is known as a *randomized, controlled trial*. You first define a group of patients, such as those who have had a heart attack but are healthy enough to undergo surgery. You then decide at random who gets treated by surgery and who serves as a *control*. This is normally done by a computer, although the process is essentially the same as flipping a coin. Finally, you observe what happens in each group and compare death rates statistically.

Randomization ensures that patients in the two groups, surgery and control, are likely to be as similar as possible. This is true both for things that you can measure (such as age, gender, or blood pressure) and also for things that are difficult or impossible to measure (such as diet and exercise, or yet to be discovered genetic factors that influence heart function). Accordingly, the only important difference between the two groups is that one received surgery and the other did not. If, at the end of the trial, there were fewer deaths in the surgery group than in controls, this would be difficult to explain other than by saying that surgery caused an improvement in survival.

One interesting point about randomized trials: it has been widely accepted since the 1950's that randomized trials are the best way of finding out which medical treatments do more good than harm. They are how we know that some things work—everything from the polio vaccine, to cancer chemotherapy, to acupuncture for pain—and other things don't, such as freezing the stomach with cold alcohol to treat ulcers (note: doctors really did use to do this). However, it wasn't until the mid-1990's that any attempts were made to collect the results of all randomized trials in one place, so that doctors and patients could access them easily. That initiative is called the Cochrane Collaboration (www.cochrane.org) and has played a critical role in ensuring that modern medicine is based on the best possible evidence.

2. Imagine that you were conducting a study on cheating at college. Like the adultery research, this involves questions about bad behavior. How would you encourage truthful answers?

I am not a psychologist so I can't tell you for sure, but it strikes me as difficult to tell someone face-to-face that you have done something that you shouldn't have. It is probably also a bit of a disincentive to truth-telling if your answer could be traced back to you (i.e., tell the truth, get thrown out of college). So here are the sorts of things I would think about in designing a study on cheating:

- Study participants should be reassured that their answers are completely confidential. They should be identified only by a code number and be told that there is no way that their answers could be traced back to them.

- Participants should answer questions using a paper questionnaire or a computer interface, rather than via an interview.

- The phrasing of questions should attempt to help participants feel comfortable about revealing that they had done something they shouldn't have (e.g., instead of "Cheating is immoral, dishonest, and destructive. Have YOU ever been GUILTY of CHEATING?" how about "Some students sometimes cheat to get better grades. Have you ever cheated on anything that contributed to your grade point average?").

CHAPTER 5

You can't have 2.6 children: On different types of data

1. **Does the median (or, say, upper quartile) always take a number that is part of the data set?**

   The median is the number "half way along" when you arrange the data from lowest to highest. If you have an even number of observations, there is no number "half way along" such that half the observations are higher and half are lower. As a trivial example, four students score 650, 690, 700 and 735 on their Math GRE. There is no number "half way along," so you take the middle of the two numbers either side of half way. The middle of 690 and 700 is 695, so the median of this data set is 695. As you can see, half of the observations are above 695 and half below. However, 695 is not in the data set.

2. **I described a continuous variable as one that can take "a lot of different values." How many different values is "a lot?"**

   Statisticians disagree on this point (statisticians disagree on a lot of points, which just goes to show how much of statistics is a judgment call). We don't go around attending statistics seminars entitled "Continuous variables: How many different values counts as enough?" where various learned professors argue over the number 6. But if you read scientific papers, you can tell that statisticians vary in their approach. For example, a very common type of data in medical research is a 0–10 scale (e.g., "rate your pain on a 0–10 scale, where 0 is no pain and 10 is the worst pain you could imagine"). Some statisticians use statistical methods suggesting that they treat the 0–10 scale as 11 separate categories. Most others use methods designed for continuous variables. Speaking for myself, I tend to like treating variables as continuous unless, as in the family size example, there are good reasons not to.

3. **Here are some variables. Which of these are continuous and which are categorical?**

   a. Height: continuous.

   b. Gender: categorical, in fact, a special type of categorical variable: there are only two categories, so gender is described as a "binary" variable.

   c. Years of education: it depends. Sometimes education is defined in terms of number of years of education (e.g., 12 for a high school graduate), in which case this is a continuous variable. However, education is sometimes put in categories (e.g., some high school; high school graduate; college graduate; post-graduate study).

   d. Pain score: pain is often measured on 0–100 or 0–10 scale, in which case it is usually treated as a continuous variable. However, participants in pain studies are sometimes asked whether they have no pain, mild pain, moderate pain or severe pain, in which case the pain variable would be categorical.

   e. Depression: again, this depends on whether depression is measured on a scale or whether patients are categorized as having no depression, minor depression or major depression.

   f. Income: Continuous, although some investigators (for reasons that are not always clear) categorize income (e.g., <\$30,000; \$30−\$49,950; \$50,000−\$74,950, etc.). This makes the point that any continuous variable can be categorized.

g. Race: categorical. Race is an interesting type of category because there is no "order" to race. If you categorize depression into none, minor and major, you have to say it in that order. It doesn't make sense to say, "We categorized depression into minor, none or major." There is similarly an order if you categorize education into some high school, high school graduate, college graduate and so on: college graduate is more education than high school graduate and high school graduate is more education than "some high school." But there is no order to race: you can say the races in order you please and it doesn't make any sense to believe that, say, Asian is any more or less than, say, Pacific Islander.

h. Unemployment rate: if you had a set of data in which a person was classified as "employed" or "unemployed," the unemployment rate wouldn't be a variable—it would be an estimate. On the other hand, you might have a data set in which each observation was a different state in the US, or a different country, with the unemployment rate given for each. In this case, the unemployment rate would be a continuous variable.

4. **Saying that an "average of 2.6 children is a silly statistic" allowed me to make some nice teaching points about different types of data. But as it happens, the "average" number of children that a woman bears over the course of her lifetime is actually pretty useful. How do you think this statistic is used?**

The average number of children that a woman bears over the course of her life is called the *fertility rate* (there are actually several different flavors of fertility rate, but I'll just use this as the simplest definition). Despite what I suggested in the chapter, fertility rate is not a simple mean: you don't just count up the number of children and divide by the number of women. This is because, for example, a woman currently aged 32 with one child may have one or more additional children in the next few years. One alternative would be just to look at women aged over, say, 45. But this would mean that we would be looking at the fertility rate of a country 20 or 30 years previously. Moreover, it isn't clear what you'd do about the fact that women live to different ages (would you sample all women or just those aged 45–55?). So the fertility rate is calculated using a complex formula, which takes into account the current number of women in various age categories, the number of children they gave birth to, and the probability that a woman survives her childbearing years.

The way in which fertility rate is used illustrates a nice distinction between two different sorts of data. Normally we think of a data set consisting of, say, people, and characteristics of those people (e.g., age and weight). In the chapter, I suggested that the data set from which "an average of 2.6 children" was calculated consisted of a group of women and the number of children they had given birth to. Such a data set might look like this:

Name	Age	Number of children
Christine	43	4
Jackie	32	1
Beth	44	2
etc.		

But you can also have data sets where each line of your table is *aggregate* data. For example, we could look at fertility rates for different countries:

Country	Fertility rate	Gross domestic product (GDP) per capita
Afghanistan	6.58	$1,000
Albania	2.02	$5,700
Algeria	1.82	$6,700
etc.		

Then we could use the data to make graphs such as the one below (helpfully provided by the CIA) (really).

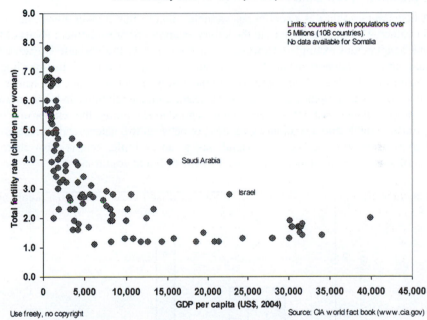

This analysis shows quite clearly that fertility rate is strongly associated with wealth, in that women in richer countries tend to have fewer children. Alternatively, we could look at fertility rates over time in just one country:

Year	Fertility rate
1970	2.43
1980	1.90
1990	1.64
2000	1.64
etc.	

Government statisticians might use these data to predict what the population might look like in, say, 2050. In many countries, the fertility rate has declined over time with the result that there will eventually be a large number of older retired people with a relatively small number of working age adults able to look after them and keep the economy going. Planning for demographic shifts of this sort is exactly the reason why statistics are such a fundamental part of government.

CHAPTER 6
Why your high school math teacher was right: How to draw a graph

1. Can you always draw a line?

A line assumes that x is a continuous variable—that is, that it takes a large number of different values. This is true for age (in the lottery example) or pre-treatment levels of headache (in the acupuncture example). However, sometimes our x, the variable we want to use to explain our y, is a categorical variable. For example, it would be difficult to plot a line showing the relationship between gender and playing the lottery, or between race and headache score. In other words, line graphs are very useful for examining the relationship between two continuous (or quantitative) variables; they are not so good at examining the relationship between a categorical variable and a continuous variable, or between two categorical variables.

This is where bar charts can be useful. Just as an example, here is a bar chart showing length of guaranteed paid maternity leave in some different countries:

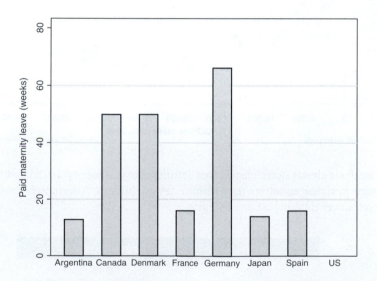

You can instantly see that there is quite a lot of variation, that some countries mandate about four months, others about a year and that the United States guarantees no paid maternity leave at all.

CHAPTER 7

Chutes and Ladders and serum hemoglobin levels: Thoughts on the normal distribution

1. Is the distribution of the results of a game of chance, such as Chutes and Ladders, really a normal distribution?

Actually no. It follows something called the binomial distribution. But if the number of observations is large, and the probability of the event (e.g., probability of winning the game) not too close to 0 or 1, then the binomial distribution ends up very close to the normal distribution. For example, you can use the binomial distribution to work out that if you toss 1000 coins, you have a 2.5225% chance of getting exactly 500 heads. If you use the normal distribution, you calculate 2.5221% instead. Now most folks wouldn't argue about 0.0004% one way or the other, and so would be happy to use the normal distribution as a decent approximation.

2. Why doesn't the graph of hemoglobin levels in the Swedish men follow a perfectly smooth curve?

Each of the bars on the histogram is a proportion. For example, take the bar at 15 gm/DL, which is close to the median; you can see from the graph that just less than 5% of the sample (the actual number is 4.7%) have a hemoglobin level of 15.0 gm/DL. Now if you did the study again, you might not see *exactly* 4.7% of men having a hemoglobin of 15.0 gm/DL, though you probably wouldn't be far off. So each of the bars on the graph is going to go up and down a little bit each time you run your study. Statisticians call this sampling variation.

CHAPTER 8

If the normal distribution is so normal, how come my data never are?

1. Can you transform all skewed distributions to a normal distribution by log transformation?

The distribution of data depends on the underlying process creating the data. In the Chutes and Ladders example, data on the number of games won and lost depends on adding a large number of random events, and so follows a normal distribution. In the prostate cancer example, which depends on a process involving multiplication, you get what is called a *log-normal* distribution. This means that the data are normal after you take logarithms.

But there are numerous other ways in which data can be generated. As an example, ability often follows what is called an *exponential* distribution, a simple example of which would be $y = 2^x$. Let's imagine that we give a bunch of people a math test and time how long it takes them to complete it. The better you are at math, the faster you'll do the test; it seems reasonable to suppose that a math whiz who is twice as good at math as the average will complete the test in half the time. If we create an exponential distribution of math ability and then calculate times as $1 \div$ ability, we get:

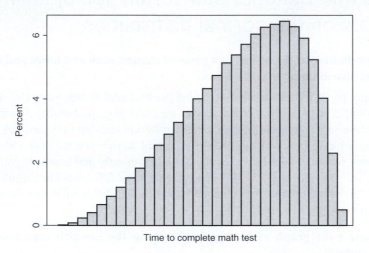

This is what we get when we log transform:

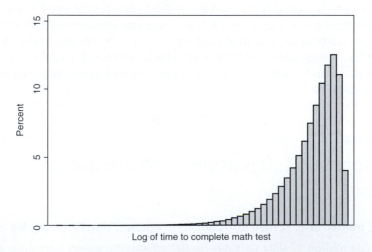

Neither graph looks anything like a normal distribution. Log transformation will only create a normal distribution if the process creating the data is log normal. For another example, have a look at the data on length of pregnancy (see *A skewed shot, a biased referee*). These data are skewed because doctors intervene and induce birth if pregnancy goes more than a few weeks past the due date. As a result, you get quite a few babies born many weeks early, but none at all born many weeks late. The process creating these data is not log normal, so log transformation will not help.

2. *For enthusiastic students only:* **At one point I said that $\log(10) = 1$. Later, I mentioned e. If you look at the graph of PSA values, you can see that a PSA of 10 comes just after the peak representing the most common PSA level. If you then look at the graph of log transformed PSA values, you can see that the most common value is around 2. So my log transformation turned 10 into a number slightly over 2, rather than 1. Why?**

Logarithms (logs for short) work like this: if $\log_x(y) = z$, then $x^z = y$. The value x is called the "base." To keep things simple to start with, we'll use base 10. The examples of logs I gave in the text were $\log(10) = 1$; $\log(100) = 2$ and $\log(1000) = 3$. This is because $10^1 = 10$, $10^2 = 100$ and $10^3 = 1000$. I used base 10 because it is easy to see things like $10^{1+2} = 10^3$, showing that logs turn multiplication into addition.

Statisticians don't like working in base 10. They prefer to do logs using base e, where e is the mathematical constant equal to 2.71828. This is because e reflects an inherent mathematical property of growth processes. The log of 10 to base e is 2.302, because $e^{2.302} = 10$.

That is the end of everything you need to know for statistics, so move on to the next chapter if you want. On the other hand, e is a really interesting number, so here are some more thoughts about it.

First, how do we get $e = 2.71828$? (Actually, the number of decimal places in e goes on forever.) The best way to think about the origin of e comes from banking. Let's imagine that you deposit $1,000 in a bank and they promise you an annual interest rate of 100% (not very realistic, but it makes things simpler to use round numbers). Let's also imagine that the bank adds the interest to your account only once, at the end of each year. So after 12 months, you'll have exactly twice as much money as when you started.

Now let's imagine that you go to the bank after six months, and demand your money back. You'll get your $1,000 plus half a year's interest, $500, for a total of $1,500. You then take your money across the street to another bank and leave it there for another six months. Six months' interest on $1,500 is $750, giving a total of $1,500 + $750 = $2,250. If you go back to the bank four times, you'd start with $1,000, then get $1,000 × 1.25 = $1,250, which would turn into $1,250 × 1.25 = $1,562.50, then $1,562.50 × 1.25 = $1,953.13 and eventually $1,953.13 × 1.25 = $2,441.41.

Now if you wanted to work out what you'd get if you took out and then reinvested your money monthly, you'd have make 12 separate calculations in a row. This is a bit time-consuming and it becomes quicker to use a formula. This is simplest when you make it work out the *growth* in your savings, so we'll call the amount you start with 1 (e.g., if you end up with 2, you have doubled your money). Also, we'll call the number of periods n. This gives

$$\text{Growth} = (1 + 1 \div n)^n$$

Applying this formula when you took money out of the bank after six months and then reinvested it, we get $(1 + 1 \div 2)^2$ which is 1.5^2 or 2.25, which is what we calculated before (your savings grew from $1,000 to $2,250, a growth rate of 2.25).

What if you calculate interest more often?

- Quarterly: $(1 + 1 \div 4)^4 = 2.441$
- Monthly: $(1 + 1 \div 12)^{12} = 2.613$
- Daily (including weekends): $(1 + 1 \div 365)^{365} = 2.714$

You may notice that the growth of our savings is getting closer and closer to $e = 2.71828$. The formal definition of e is $(1 + 1 \div n)^n$, where n approaches infinity. For example, if we plug $n = 1,000,000$ into the formula, we get 2.7182805. Using n of 1,000,000, means we are dividing growth into 1,000,000 separate periods.

So e describes growth, where time is divided up into a very large number of periods. This is the same as assuming that growth happens constantly. (Which it does, right?) As result, we find *e* popping up all over the natural world: spirals (which describe the growth of snail shells), radioactivity (decay is the opposite of growth), how far ultrasound penetrates through body tissue (you can think of the strength of a signal as decaying), the shape of a rope hanging between two points (the effect of gravity grows the farther you get from where the rope is fixed), an arch under a bridge (which is sort of the opposite of a hanging rope). The formula for the normal distribution also includes *e,* which means that *e* helps describe an enormous number of natural phenomena (such as hemoglobin in middle-aged Swedish men—see *Chutes and Ladders and serum hemoglobin levels: Thoughts on the normal distribution*), as well as the results of repeated experiments. The constant *e* is also found in Euler's identity, which is $e^{i\pi} + 1 = 0$. The remarkable thing about this equation is that it links three very different mathematical constants, *e,* to do with growth; *i,* the square root of -1; and π, the circumference of a circle divided by its diameter, along with "+1" which is how you get all numbers, and zero, a key mathematical concept. Mathematicians like to trip out over Euler's identity, seeing it as a demonstration of the deep mathematical harmony of the universe. I am not sure I'd go that far, but it certainly speaks to something profound.

CHAPTER 9
But I like that sweater: What amount of fit is a "good enough" fit?

1. Isn't statistics meant to be very precise? Don't we prefer "28.29%" to "about one in three?"

Giving results to many decimal places certainly *sounds* very scientific. This is perhaps best seen in magazines like *Cosmopolitan* or *Men's Health*, which tend to report that, for example, "48.2% of men think that women should offer to pay their share on a first date." Somehow that ".2%" gives the statement an air of scientific seriousness not afforded by stating that "about half" of men like women to pony up. The point is, of course, that "about half" is all we need to know. It makes absolutely no difference to any opinion that we'd hold or action we'd take whether the true proportion was 48.2%, or 50.0% or even 56.7%.

The other problem with "48.2%" is that, if the survey was repeated tomorrow, you'd probably be surprised if *exactly* 48.2% of men responded "yes" to the question about splitting the bill. It might be 53.9% or 45.7% (although probably not 86.5%). So "48.2%" suggests a level of precision that you just don't have.

One final point about those magazine surveys: they rarely if ever use random sampling. One recent survey I saw ("our results reveal who the modern man is") reported findings such as "56% of men see the drink they order as a reflection of their masculinity or character." The survey was conducted via a click-through on a website and thus excluded: (a) men who don't use the Internet; (b) men who speak poor English; and (c) those of us with better things to do than fill out manliness surveys online. So it isn't so much "56% of men have hang-ups about their choice of alcoholic beverage" but "56% of men who happened to be on our website were intrigued by a picture of an attractive, under-dressed female and then had the patience to click all the buttons on our questionnaire."

2. The histogram showing test scores is skewed to the right. Why would that be?

The figure gives the distribution of a measure of ability. Measures of ability are often skewed to the right because people like doing what they are good at (the jock works out, the geek programs computers), and practice makes them even better. Also, the more knowledge you have, the easier it is to gain more (it is easier to learn Italian if you speak French, English, Spanish and German than if you speak only English). As a result, you generally have more very high achievers than you'd expect from a normal distribution.

3. **It is generally said that "95% of observations are within two standard deviations of the mean." To calculate where 95% of the observations were for the table, I multiplied the standard deviation by 1.96 rather than 2. How come?**

The idea that "95% of observations are within two standard deviations of the mean" is just a rule of thumb. To get the actual number of observations within two standard deviations of the mean of a normal distribution, you have to do some complex math. If you plug the numbers into the appropriate formulas, you find that actually 95.45% of observations are within two standard deviations of the mean and that 95% are within 1.96 standard deviations. So you use "two standard deviations" to check things quickly by eye, and "1.96 standard deviations" if you are doing calculations.

CHAPTER 10
Long hair: A standard error of the older male

1. **What I am describing as a "study" here is when I go to a single party and measure the hair length of every guy. The aim of the study is to estimate the typical hair length of American men. What must I assume in order to use my data (hair length at one party) to inform my study aim (hair length of American men)?**

My main assumption is that guys attending the sort of parties I go to in New York are a random sample of all American men. This is a pretty questionable assumption, because different parties attract different sorts of guys (and some guys never go out at all). The mean hair length at a graduation ceremony for US Marines would trend low; guests at *The Sixties Live On* party in San Francisco would probably finger me as a cop. This is similar to the opinion poll that famously made an entirely incorrect prediction for the results of the 1936 presidential election (see *A skewed shot, a biased referee*). The problem with the poll was that the people surveyed in the poll were wealthier than average, and wealthier people had different voting patterns from the population as a whole. Random sampling is an absolutely key idea in statistics: if your sample isn't random, then your estimate will often be biased.

Incidentally, if you want a quick definition of random sampling, try this. Random means "unpredictable," so imagine that two people walk into the room, one who was sampled and one who wasn't, and you have to guess which one of the two was sampled. If you have no better than a 50% chance of guessing correctly, then you have a random sample. In the case of the presidential poll, you'd ask the two individuals about their finances and guess that the richer one was the one in the sample: you'd be right more than 50% of the time, showing that random sampling was not used.

2. **What has statistics got to do with parties?**

As it happens, one of the reasons that modern statistics got going at all was . . . beer. One of the most commonly used types of statistical test is the *t* test. This was developed by William Sealy Gosset, who worked at the Guinness brewery in Dublin and developed statistical methods to help with quality control for beer making (Gosset wrote under the name "Student," which is why the *t* test is sometimes referred to as "Student's *t* test"). R A Fisher, another of the founders of modern statistics, conducted much of his early work on barley, for example, analyzing experiments to see which strain of barley grew fastest. The main use of barley is to make beer.

CHAPTER 11
How to avoid a rainy wedding: Variation and confidence intervals

1. **When we were trying to guess the body mass index of the student, I stated that 95% of the individual observations would be within about two standard deviations of the mean. I said something similar about reference ranges in the "Things to Remember." What is my assumption here?**

   Statements about what proportion of observations are within a certain number of standard deviations of the mean depend on the assumption that the data are approximately normally distributed (although see *But I like that sweater: What amount of fit is a "good enough" fit?*). It turns out that once we get rid of the athletes (including the 300 lb offensive lineman), the body mass index of male students does follow a normal distribution (or at least it did in the data set I looked at). This is somewhat different to the distribution of body mass index in adults (see *A skewed shot, a biased referee*), which is very skewed.

2. **When talking about the results of the lecturer's study on body mass index, I said that "95% of study results—the mean BMI—will be within two standard errors of the mean." What is the mean here? What is the standard error?**

   The mean is the true mean; the standard error is the true standard deviation, divided by the square root of the sample size. The true mean and true standard deviation are what are known as *parameters*. The lecturer had a data set of 100 BMI's from men at the college who didn't play sports. Statisticians assume that these BMI's were randomly selected from an infinitely large group of theoretical non-athlete male students. This population has a certain mean and standard deviation, which are the population parameters. The mean and standard deviation we calculate for the weight of the 100 students in our sample are estimates of these parameters.

   In a typical analysis, you can never know the true values of population parameters such as a mean or standard deviation. So saying that "95% of means are within two standard errors" is a theoretical point. It is easy to demonstrate using computer simulation: I tell the software what the true mean and standard deviation are; simulate a large number of studies; calculate the mean of each study; calculate the standard deviation of the study means (because this is the standard deviation of study results, we call it the standard error); show that 95% of the study means are within about two standard errors of the true mean.

   Because you don't know the true population mean and standard deviation, you have to be a little careful when interpreting the results of a study. Imagine that you did a study where you measured 100 students and calculated a mean BMI of 26.5 and a standard deviation of 2.5. You could then calculate the standard error as 0.25 (to get standard error, you divide the standard deviation by the square root of the sample size) and calculate a 95% confidence interval of $26.5 \pm 0.25 \times 2 = 26.0$ to 27.0. What you *can't* then say is that, "If I repeated my study a large number of times, 95% of the time, the mean body mass index would be between 26 and 27." This is because you aren't using the true mean and true standard deviation, but the estimates you obtained from your study. Statisticians tend to define confidence intervals by saying, "95% of 95% confidence intervals include the true mean." It is a subtle point, and probably not worth worrying about too much: just think about the confidence interval in terms of a range of plausible values for your study results.

CHAPTER 12
Statistical ties, and why you shouldn't wear one: More on confidence intervals

1. **I argued that, if we had to bet on it, we should put money on the Democrat, even though the confidence interval for the poll included the possibility that the Republican would win. Does this mean that we should just abandon confidence intervals then and go with whatever looks best?**

   The key phrase here is "if we *had* to bet on it." If you are forced to make a choice between two alternatives, you should go for the one likely to give you the best result, pretty much regardless of how much better it is, or how sure you are about your choice. As an example, imagine that you are a soldier and that you have been captured by the enemy and thrown into prison. You manage to escape your cell and make it through the prison until you get to two doors, one of which leads outside to freedom and the other to the guard's room and certain death. Now you happen to remember a study showing that 51% of guard room doors smell of beer, whereas only 50% of exit doors do so. Now although this isn't much to go on, you're better off choosing the door with the less beer-like smell than just flipping a coin.

   So when you are forced to make a decision between two similar alternatives, you ignore the confidence interval. It follows then that if you aren't forced to make a decision, or the alternatives aren't similar, the confidence interval can be very useful indeed.

   Firstly, you don't always have to make a decision straightaway. Imagine that you worked for a computer company and had conducted a study to see which of two types of track pad laptop users prefer: "original" or "new." Of the 50 users you study, 30 (60%) prefer the new track pad. The confidence interval for 60% of 50 is 45% to 74%. Although it looks as though users prefer the new track pad, it may be that more people in fact prefer the original version. When you report these findings to your boss, she tells you that production of the latest model of the laptop is still some months away, and you don't have to make a decision yet. Indeed, given that there are many millions of dollars at stake, it would seem worth getting more data before making a final recommendation. When you repeat your study on a further 50 users, you find that 35 prefer the new track pad. This gives a total of $30 + 35 = 65$ in favor out of $50 + 50 = 100$, in other words, 65% favorable with a confidence interval of 55% to 74%. It is now unlikely that most users would prefer the original track pad and you can now recommend that the new track pad go into production. This is an example of how increasing the sample size can decrease the width of the confidence interval. It turns out that you have to be a little careful if you just keep adding more observations until you get a statistically significant result, but the general point remains: if you don't have enough evidence, and you don't have to make a decision yet, get more evidence and then make a decision.

The other reason why a confidence interval can be useful is when the two alternative decisions are not similar. In the track pad example, it may be that switching production to the new track pad would be expensive, and so everyone wants to be pretty sure that users really do prefer it. Alternatively, it might be that the original version of the track pad has been around a long time and has been very popular, and the new track pad is a pretty major design departure. Your boss would be justified in saying something like, "Look, our laptops are very popular, and we are not going to push through a major design change unless we are really sure our potential customers are going to prefer it." Another example comes from my own field, medical research. In a clinical trial of a new drug, we are usually very interested in the confidence interval. If the confidence interval includes the possibility that the drug could actually be worse than placebo, we say that the trial was "negative," because you don't start prescribing a new drug until you have pretty firm evidence that it is of benefit.

2. **Opinion pollsters for political races typically survey around 1000 people, but they don't go out in the morning and ask the first 1000 people they meet. What do they do?**

If your first answer was "conduct a random sample," that is a good start. People out and about on the street in the morning are not a representative sample of voters: you'd probably get a lot of parents with young children, shift workers and retirees, and it is unlikely that these groups vote exactly the same as people who work in the morning.

As it happens, opinion pollsters don't usually try too hard to get a random sample. In fact, some would argue that a random sample isn't even theoretically possible for opinion polls about politics. If I wanted, say, a random sample of lawyers in New York State, I could presumably get a list from the New York Bar Association and then randomly select some to call. Alternatively, if I wanted a random sample of fans at Yankee games, I could go to Yankee stadium and approach, say, 10 randomly selected people in each section of seating. However, if I am conducting an opinion poll, I can't randomly select voters, because the election hasn't happened yet: we know who is a lawyer, and we know who is at a Yankee game; we don't know who will vote in a forthcoming election.

What opinion pollsters do instead of random sampling is to use "weighted" sampling. To give an example (with some silly numbers to make the math easy), let's say that an opinion poll included 600 registered voters, two-thirds men and one-third women. The 400 men went 300 to 100 for the Republican; the 200 women went 125 to 75 for the Democrat. This gives a total of 375 Republican to 225 Democrat, a 62.5% to 37.5% victory. However, imagine that in the last election for the Senate, two-thirds of those who voted were women and one-third were men. To make our sample look similar to that voting in the previous election, we have to count each woman twice (a "weight" of 2) so that they make up two-thirds of the electorate and count each man as a half (a "weight" of 0.5) so that they make up one-third. If you do the math you get 50% voting Democrat and 50% voting Republican (a real tie, not a statistical one). In practice, pollsters use weighting schemes that are quite a bit more sophisticated, including gender, age, party affiliation and past voting history.

CHAPTER 13
Choosing a route to cycle home: What p-values do for us

1. **I stated that I "provided strong evidence that using the busy road was quickest." Why didn't I just say that I'd proved it?**

   "Proof" is not a word often used by scientists. Outside of the movies, few scientists will announce, "Aha! I have proved my theory." (Also note that, outside of the movies, few scientists have crazily unkempt grey hair, or run experiments with test tubes containing bubbling, bright red fluids.) Statisticians are particularly careful with the word "proof," because they are keenly aware of the limitations of data, and the important role that chance plays in any set of results. Statisticians normally use the word "proof" only to refer to mathematical relationships between formulas. The point here is that you don't use data to do math theory, so you aren't subject to the limitations of data, and so can go about really claiming to have "proved" something. It is certainly unwise to think that you can prove something by applying a statistical test to a data set.

2. **We normally think that a big difference between groups means a small p-value. But I found a very small p-value ($p = 0.001$) even though the difference in travel times between the two different routes home was trivial. How come?**

   The p-value is about the strength of evidence; a low p-value means stronger evidence. Now one way that you might have strong evidence is if you have a big difference: if the mean travel time on the backstreets was an hour vs. 5 minutes on the busy road, we'd see this as a pretty firm demonstration that the busy road was quicker, and we'd expect a low p-value. But we also consider evidence to be strong if there is a lot of it. For example, if we asked a million men and a million women about their opinion of the president, we'd probably be comfortable in saying that any difference, pretty much no matter how small, represented a true difference between the sexes (as it happens, even a 0.2% difference would be statistically significant for a sample size this large). I obtained the p-value of 0.001 for the difference in travel times by looking at two years of data, about 450 trips in total. This is a lot of trips, which is why a 57 second difference, though small, was statistically significant.

   So the p-value depends on both the size of the effect and on the number of observations. This is one reason why you can't look at a p-value and decide whether something is important or not. An effect might exist; whether you should pay it any mind depends on whether it is large or small. This is why we normally need estimation as well as inference.

3. **If statistics is not just about testing hypotheses, what else can you use statistics for?**

   Testing hypotheses is *inference*. The other thing statisticians do when they analyze data is *estimation*. When I report the results of my two-year study of travel times I found "strong evidence that going home via the busy road is faster ($p = 0.001$), but not by much (it saves me 57 seconds on average)." The p-value of 0.001 allows us to make an inference about which way home would be faster; the "57 seconds quicker" on the main road is the *estimate*.

CHAPTER 14

The probability of a dry toothbrush: What is a *p*-value anyway?

1. **Why do we say the probability of the data under the null hypothesis? Wouldn't it be more interesting to know the probability of the hypothesis given the data?**

The short answer is yes, it would be great to know the probability that the hypothesis was true. But no, unfortunately, traditional statistical tests don't tell you that. One key point is that when you analyze a data set, you don't incorporate any information from outside your study. Sometimes there are very good reasons (other than your study) to believe that a hypothesis is either true or false. I was recently shown a study which suggested that a certain type of kidney cancer was more likely to be fatal than another type ($p = 0.04$). My first comment was, "Didn't we already know that?" A large number of studies had already been published all showing the same thing and, if I remember correctly, there were also some animal studies examining exactly why it was that certain types of kidney tumor were more aggressive. On the other hand, there is an idea that all cancers are caused by a parasitic infection and can be cured by a special "zapper". (You can't make this stuff up.) If you showed me a medical study showing that these zappers cured cancer with a *p*-value of 0.04, I'd probably say something like, "Well, that is surprising, but it is a ridiculous hypothesis, and there is no reason to believe it is true other than this one measly *p*-value. So thanks but no thanks, I am not going to believe in this hypothesis for now." I'd probably also look to see if the study was well conducted. A statistical analysis can't tell you the probability that a hypothesis is true, because data cannot distinguish between a good study and a flawed one.

There is another reason why we say "probability of the data if the null hypothesis were true" not "probability that the null hypothesis is true given the data." This goes back to how we work out the math behind statistical tests in the first place. It is possible to calculate the probability of a particular result if a certain hypothesis were true. A trivial example would be the hypothesis that "this is an unbiased coin" for an experiment where we toss a coin four times. We can calculate, for example, that if the hypothesis were true, the probability of throwing four heads in a row would be $1 \div 2^4 = 6.25\%$ and the probability of throwing three heads out of four would be $1 \div 2^4 \times 4 = 25\%$. We can do something similar for continuous variables: for example, under the hypothesis that "the mean height of male athletes is 6 ft," there is a 50% probability that the mean of a sample of male athletes is 6 ft or more.

However, it is not possible to calculate the probability of a hypothesis given that we have a particular result. Given four heads in a row, what is the chance that the coin is unbiased? Given a mean height of 6 ft 2 in., what is the probability that the true mean height is 6 ft? It is unclear how we could work this out. The problem is that we have an infinite number of possible hypotheses (e.g., the coin is unbiased, the coin is biased 60:40, the coin is biased 69.999999:30.000001 etc.; the mean height is 6 ft, the mean height is 6 ft 0.001 in. the mean height is 2000 ft and so on).

Incidentally, there is a special branch of statistics, Bayesian statistics, that does try to estimate the probability of hypotheses. Roughly speaking, Bayesian statisticians start by stating a subjective probability of a hypothesis before the data from a study are made available. They then update this *prior* probability depending on the data: negative data make the hypothesis less likely; positive data make the hypothesis more likely. Bayesian methods remain somewhat a minority pursuit in most areas of statistics, at least partly because of the difficulty of specifying the prior probability of a hypothesis.

2. **Here are some research questions. Give an example of the null hypothesis for each of these:**

   a. Does compulsory job retraining affect long-term unemployment? Example null hypothesis: *There is no difference in rates of long-term employment before and after job training was made compulsory for welfare recipients.*

   b. Do African American males have a harder time than white males hailing a taxi in New York City? Example null hypothesis: *The time to hail a cab is the same for African American and White males.*

   c. Nationwide, about 28% of births are via Cesarean delivery. Do hospitals in New York State have higher than average Cesarean rates? Example null hypothesis: *The Cesarean section rate in New York State is 28%.*

   d. Do after-school programs increase student participation in art, music, drama or dance activities? Example null hypothesis: *There is no difference in student participation in art, music, drama and dance between schools with and without after-school programs.*

   e. Do patients taking a new, less toxic type of chemotherapy have response rates at least as good as those on the standard (and unpleasant) chemotherapy drug? Example null hypothesis: *Response rates on standard chemotherapy are higher than those on the new chemotherapy agent.*

A couple of comments: first, the null hypotheses above are just examples and often several different null hypotheses are possible for any particular question. For example, for question (d), you might also have the null hypothesis that "There is no difference in student participation in art, music, drama, and dance comparing the period before the introduction of school programs to New York City schools to the period after they were introduced." Alternatively, what about "Children offered a place in an after-school program have the same level of participation in art, music, drama, and dance as children not offered a place." These three hypotheses reflect different study designs, respectively, comparing different schools during the same period of time; comparing different periods of time in the same schools; comparing different students in the same schools during the same period of time. The link between study hypothesis and study design is a central tenet of research methodology.

The second comment is that the null hypothesis for (e) is that there *is* a difference between groups. This is odd because the null hypothesis is often defined in terms of "no difference between groups." The reason why we have a null hypothesis that standard cancer treatment is better than the new drug is that, if true, nothing would change, we'd just keep doing what we normally do and give the usual treatment. This is a somewhat unusual case, so don't worry too much if you can't quite wrap your head around it. I raise it because it shows that the null hypothesis is pretty hard to pin down. Perhaps we should expect this: we apply inference statistics to all sorts of areas of science, so we need null hypotheses for all sorts of different scientific questions; it would be surprising if we could define all possible scientific questions in a single sentence.

CHAPTER 15

Michael Jordan won't accept the null hypothesis: How to interpret high p-values

1. **When discussing statistical significance, I have repeatedly described p-values less than 0.05 as "statistically significant" and p-values of 0.05 or more as "not statistically significant." Is it true that a p-value of 0.049999 is always statistically significant and that a p-value of 0.050001 is never statistically significant?**

   The most commonly used threshold for statistical significance is 0.05. This is why I have referred to p-values < 0.05 as "statistically significant" and p-values of 0.05 or more as "non-significant." However, in theory you can choose whatever level you want to determine statistical significance. This level is called α (*alpha*). If you set an α of 0.01, you call p-values < 0.01 statistically significant. Accordingly, a p-value of 0.04999 would not allow you to reject the null hypothesis if your α were 0.01. Comparably, if your α was 0.1, a p-value of 0.050001 would be statistically significant and would lead to rejection of the null hypothesis. Saying that "the most commonly used threshold for statistical significance is 0.05" is therefore the same as saying "the most commonly used value of α is 0.05." In theory, you could set an α of, say, 0.03869, but this is never done in practice.

   Incidentally, it is important to set your α *before* you run your analysis. I have sometimes seen results such as "The p-value for the difference between groups was 0.0684. Our results are therefore statistically significant at the 0.07 level." In short, if you set α after you get your p-value, you can ensure that your results are always statistically significant. Of course, this is great news if you are, say, a genetics company trying to sell a useless test; the rest of us, however, might prefer it if everyone just stuck to 0.05.

2. **Why do you think that my experiment with Michael Jordan resulted in a non-significant p-value?**

   Assuming that you did a fair study—and that was the case here, it wasn't as if I was blindfolded or anything—there are two reasons why you can get a non-significant p-value. The first possibility, obviously enough, is that the null hypothesis is true. The second possibility is that although the null hypothesis is false, you don't yet have enough evidence to reject it.

   One way of thinking about the p-value is that it represents the strength of evidence against the null hypothesis. If you have a high p-value, and you believe that the null hypothesis is false, one option is to go and collect more evidence. In the case of the basketball experiment, that just means asking both me and Michael Jordan to shoot some more free throws. If, for example, we both threw another 7 and I again hit 3, and he hit 6, that would give a total score of Jordan: 13 of 14 vs. Vickers: 6 of 14. This difference is statistically significant ($p = 0.013$).

   A statistician might explain this by saying that small sample sizes often result in high p-values, irrespective of whether the null hypothesis is true or false. Indeed, an important part of a statistician's work is working out exactly how big a study needs to be in order to have a good chance of rejecting a false null hypothesis (see also *Meeting up with friends: On sample size, precision and statistical power*).

3. **What should we conclude about the effects of a low fat diet on breast cancer?**

   My own view on the low fat diet trial is that the results were very encouraging, on the grounds that a 10% reduction in cancer risk is pretty important. However, the results weren't quite strong enough for us to be really confident about the effects of diet on breast cancer: it could be that fat has little effect or is perhaps even protective. Accordingly, we really need more data to be sure one way or another. As it happens, we can get more data without doing another trial, just by waiting around—more women will develop breast cancer over time and we can go back to see whether these women were assigned to the low fat or standard diet group in the trial. My guess is that the scientists running the trial will update their results in the next few years once these data have been collected.

4. **What is the connection between a criminal trial and a *p*-value?**

    Say that someone is accused of stealing a car, and the case goes to trial. The jury can decide on only one of two verdicts: "guilty" or "not guilty." Roughly speaking, a guilty verdict means that "we, the jury, find beyond a reasonable doubt that the defendant stole the car." A verdict of "not guilty" means "we, the jury, do not find beyond a reasonable doubt that the defendant stole the car." There is no verdict of "innocent."

    Statistical hypothesis testing is very similar: just as you say "guilty" and "not guilty" for the criminal trial, you say, "Reject the null hypothesis." and "Don't reject the null hypothesis." for statistical testing. And just as you can't say, "The jury found him not guilty, so he is innocent." you also can't say, "$p \geq 0.05$, null hypothesis not rejected, so the null hypothesis is true."

CHAPTER 16
The difference between sports and business: Thoughts on the *t* test and the Wilcoxon test

1. **You may have heard that statistical tests come in one of two flavors: parametric and non-parametric. The *t* test is a parametric test; the Wilcoxon is non-parametric. What does "parametric" mean and why is the *t* test, but not Wilcoxon, parametric?**

    Parametric statistics are defined as methods that assume that the study data are drawn from a theoretical distribution with certain characteristics. These characteristics are known as *parameters*; means, standard deviations and proportions are all examples of parameters.

    Imagine that we have a data set consisting of the weight of 100 sheep, and we are considering a parametric statistical analysis. What we are assuming is that the weights of these sheep were randomly drawn from an infinitely large group of theoretical sheep. This population has a certain mean and standard deviation, which are the population parameters. The mean and standard deviation we calculate for the weight of the 100 sheep in our sample are estimates of these parameters.

    As regards our cycling study, to get a *p*-value using a *t* test, we first calculated an estimate for the difference between group means and then divided this estimate by its standard error. The formula for the standard error includes the standard deviations of both the massage and control group. So to conduct a *t* test we need to calculate means and standard deviations. These are estimates of the parameters of a theoretical distribution of massaged and non-massaged student cyclists. In contrast, to get the *p*-value from a Wilcoxon, we just added up ranks, and there was no need to calculate estimates of any parameters. We might choose to report a median when we present a *p*-value from a Wilcoxon test, but we don't have to calculate the median to run the test. Hence the *t* test is parametric and Wilcoxon is non-parametric.

    A simple way of thinking about parametric statistics is shown in the following figure:

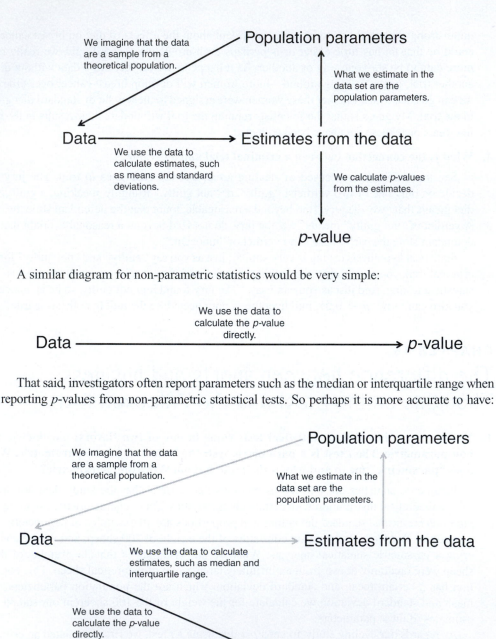

A similar diagram for non-parametric statistics would be very simple:

That said, investigators often report parameters such as the median or interquartile range when reporting p-values from non-parametric statistical tests. So perhaps it is more accurate to have:

I put the lines for the top half of the diagram in gray, because they are not an inherent or necessary part of non-parametric statistical tests.

2. **What would you conclude from our experiment about the value of massage for recovery from cycling?**

We certainly shouldn't conclude that massage improves cycling times because the p-value was well above 0.05 and, as a result, we can't reject the null hypothesis of no effect of massage. On the other hand, we wouldn't want to say that massage doesn't work because that would be to accept the null hypothesis, which you shouldn't do (see *Michael Jordan won't accept the null hypothesis: How to interpret high* p-*values*). The key thing is to look at our main estimate and its confidence interval. We estimated that massage reduces race time by about a minute, but it could reduce race times by up to five minutes. Now cyclists spend a fortune on buying lighter equipment in order to shave at most a few seconds off their time—a difference of even one minute is enormous, let alone five minutes. So what we should conclude is that the study wasn't precise enough to give us an answer one way or the other and additional research is required. The reason why we didn't get a precise result isn't hard to work out: the study was very small and small studies typically lead to wide confidence intervals.

CHAPTER 17

Meeting up with friends: On sample size, precision and statistical power

1. **When calculating the sample size needed for a study with a hypothesis test, you need to specify an effect size, the difference from the null that you want to find. How do you choose an effect size?**

    Choose carefully, because small differences in the effect size lead to big differences in the sample size. The inverse square rule applies: halve the effect size you want to try to find and you quadruple the sample size you'll need.

    One common mistake is to fix the effect size in terms of what you *expect* to see. Take the study of the new drug for the common cold. Imagine that the researchers did a preliminary pilot study on 20 patients, all taking the drug, and found that 17 (85%) recovered within 48 hours. The temptation might be to design a trial that assumed a 50% recovery rate in the control group and an expected 85% recovery rate in the drug group. Such a trial would require about 80 patients. But let's say that the results of this trial were that 28 of 40 (70%) in the drug group recovered versus 20 of 40 (50%) controls. Although this looks great for the drug— most of us would be very interested in improving our chances of an early recovery from 50% to 70%—the result is non-significant.

    This suggests that the effect size should be determined by considering the minimum effect we'd be interested in. For example:

    - A doctor might tell you that, given the costs and side effects of the new drug for the common cold, it would really need to improve recovery rates to at least 65%. If recovery rates improved to only 60%, it probably wouldn't be worth taking.

    - A psychologist interested in gender differences in learning might specify that, if girls and boys differed by less than 10 points on a test, you'd say that they pretty much had equal scores.

    - A sociologist might analyze whether conservatives and liberals give different answers to a questionnaire examining attitudes to politicians. The sociologist might state that differences of 3 points or more would be interesting.

    Accordingly, the doctor, psychologist and sociologist would design their trials with effect sizes of 15%, 10 points and 3 points respectively.

2. **Something that often happens is that a study fails to reject the null hypothesis. This typically leads the investigators to start running around to find someone to blame for this "negative" result. A common question is: what was the power of the study anyway? Is this a sensible question to ask?**

Not in my view, because power is something you think about when designing a study, before you have any results. The power is the probability that you'll correctly reject the null hypothesis. Now that the study is over, you know that probability: it is zero, because you didn't reject the null. Asking, "What was the power of the study we just analyzed?" is a bit like asking, "Do you think the Mets have a good chance against Atlanta?" the day after the Mets lose game 7 following a wild pitch in the top of the 9th.

What you should focus on at the end of the study is the confidence interval. If the confidence interval includes results that would be interesting, then it might be worth doing further research. And that is really what you should want to know, right? As a psychotherapist might put it: it doesn't matter who was to blame for the past, it is what you do about it in the future that matters.

As a quick example: in the trial of the cold drug, we stated that if the drug increased response rates by about 15% it would be worth using. Here are some possible trial results, all of which show no statistically significant difference between groups:

Trial	Total number of patients	Recovery rate in the control group	Recovery rate in the drug group	Difference between groups	95% confidence interval
a	360	50%	50%	0%	−10% to 10%
b	200	60%	50%	−10%	−24% to 4%
c	480	50%	58%	8%	−1% to 17%
d	100	50%	62%	12%	−7% to 31%

From the results of trials a and b, we'd conclude that the drug is unlikely to be worth using because the upper bound of the confidence interval—our estimate of what the best possible effects of the drug could be—is less than 15%. Trials c and d might encourage us to give the drug a second shot because the confidence interval includes the possibility that it could increase recovery rates by 15%, something which we'd consider useful.

3. **Should you always do a sample size calculation when planning a study?**

You might be surprised that the answer is actually "no." Speaking in purely scientific terms, a study just keeps getting better and better as the sample size rises: power increases and the confidence interval gets narrower. So, the ideal trial of the cold drug would include every person with a cold in the world, and the ideal study of political attitudes would give a questionnaire to every liberal and conservative in the US. The reason why we don't typically do studies with sample sizes in the millions is that they cost too much and take too long.

Moreover, studies that are too large are ethically questionable. My own field of cancer research provides a good example. From a strictly scientific viewpoint, a trial of a new chemotherapy drug should be as large as possible in order to give a precise estimate of its effects. However, every additional patient put on a chemotherapy trial costs you something. If the drug is ineffective, you have put a patient through the misery of chemotherapy for no good reason; if the drug does help fight cancer, then you have delayed the date when you can announce the results of the study (because you have to wait for the new patient's results to come in), thereby denying perhaps hundreds of patients throughout the world the opportunity to get an effective treatment before their cancer has spread too far.

The point of formal sample size calculation is to balance benefits and harms. On the one hand, the scientific benefits of having a precise result from a large sample size. On the other hand, the cost of a large study, in terms of time, money and lost opportunities. The reason why sample size calculation is not always necessary is that, for some studies, increasing the sample size doesn't incur any significant costs. If you are analyzing data that have already been collected, it doesn't cost any more time or money to download an entire data set or only part of it. This is why many economic studies don't need sample size calculations: data for things like interest rates, unemployment or inflation are already available.

Here is another example of where there was no drawback to increasing sample size. Some medical researchers wanted to test some hair samples that had been collected from AIDS patients and stored in a hospital refrigerator. The researchers hypothesized that trace levels of anti-AIDS drugs would be detectable in the hair and might predict how long the patients survived. Now although it is time consuming and expensive to set up a lab to test the hair, testing any individual hair sample is pretty quick and costs virtually nothing. Moreover, the test doesn't damage the hair so the sample can be put back in the refrigerator and studied again later on by someone else. The researchers therefore proposed to the hospital ethics committee that they test all 500 or so hair samples that had been stored. When the ethics committee refused their request and demanded a sample size calculation, the researchers pointed out that there was no downside to testing all the samples: it didn't cost much more or take importantly longer; no patient could be harmed; no research material would be lost to future scientists. All too predictably, the ethics committee again denied their request due the lack of a sample size calculation.

This raises a nice general point about statistics: statistics is a set of tools to help you find things out, not a set of rules you have to follow. Which statistical tool you use—or even whether you use one at all—depends on what it is you want to find out. The researchers were absolutely right to say that sample size calculation was not a tool that they needed; the ethical committee was wrong to believe that statistics is a set of rules, one of which states that "sample size calculations must always be conducted."

4. *For enthusiastic students only:* **When we looked at the power of our survey study, I showed a distribution for the upper bound of the 95% confidence interval. I said that when this bound was less than 50% (our value for the null hypothesis), the result would be statistically significant. As such, I claimed that this distribution was the same as that for statistically significant results. Is this actually true?**

Not quite, although it doesn't make much practical difference in this particular case. The null hypothesis in the survey study is that the proportion of students in favor of the change in exams is 50%. We reject the null if the observed proportion in our study is significantly higher or lower than 50%. This means that we reject the null if:

- The upper bound of the 95% confidence interval is below 50%, *or*:
- The lower bound of the 95% confidence interval is above 50%

In my diagram, I show only the distribution for the upper bound, but say that the "proportion of results less than 50%" is "the proportion of statistically significant results." Technically, that proportion should also include times when the lower bound was above 50%. However, this would be extremely rare for this particular survey study. For example, the lower bound of a 95% confidence interval would be above 50% if at least 61 of 100 survey respondents were in favor of the new exams. The probability of observing 61 of 100 in favor of the new exams if the true rate in favor were 38% is about 1 in 350,000.

CHAPTER 18

When to visit Chicago: About linear and logistic regression

1. **In reality, no one tries to predict marathon times in terms of age, gender and training miles. What would you use instead?**

   You might know that most marathon runners have a pretty good idea of what time they are going to run, and that this isn't based on their age, gender or training miles. Many runners have run marathons before and use their previous times as a baseline; they might then adjust up or down depending on how training is going or whether, for example, they have had an injury. Runners attempting their first marathon have normally timed themselves over shorter distances (such as a half marathon) and can make a pretty good guess on that basis (such as by doubling their half marathon time and adding 15 minutes or so).

   This makes an important point: if we really want to predict something, we have to think hard about what would be the best predictors. Choice of predictors is a subject of intense statistical debate, with different schools of thought. For example, should you specify your predictors before looking at the data, or let the data guide your choice of predictors? (I tend toward the former.) The marathon running example illustrates something else about prediction: generally, it isn't *what* you are that makes a big difference but *how* you are. For the marathon, it isn't age or gender that matters as much as current running time; in studies of pain, what predicts a patient's pain level a year from now is not their age, or type of pain or even type of treatment, but their current level of pain; in cancer, what makes the biggest difference to survival is how far the cancer has spread before treatment; if you want to guess someone's college grade point average, just look at their scores during high school.

2. **In a regression such as $y = b_1x_1 + b_2x_2 + c$, the c is called the intercept or constant. Why?**

   I like to use the term *constant* for c because it is the amount added to everybody's score, regardless of their values of the x's. In the marathon running example, the constant was 262 minutes: this amount is added no matter what the person's age, gender or training regimen. You can also use the term *intercept* for c because c is where a regression line crosses the y axis when x is zero.

   For example, take the following regression equation: $y = -0.062x^2 + 1.8x + 2.6$. Because the equation includes an x^2, it is non-linear (i.e., a curve) as shown in the following graph:

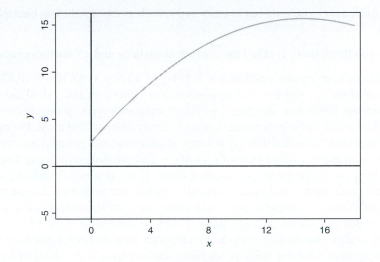

The intercept here is 2.6 and you can see that the regression line crosses the x axis at $y = 2.6$.

This next bit is for enthusiastic students only, so if you are not that interested in regression constants, skip ahead to the next answer. The complication is that you sometimes have to be a little bit careful about the intercept. What if I told you that the data for the graph were from a study of diet, where y was weight loss and x was number of weeks on the diet? The problem with the regression line is that it suggests that participants immediately lose 2.6 lbs the instant they sign up for the diet. A more sensible regression line is $y = -0.060x^2 + 2.1x$, which gives:

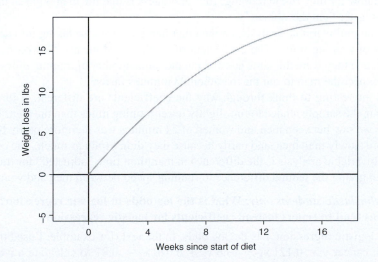

I obtained this regression equation by forcing the constant to be zero. There is an active debate among statisticians about the sort of circumstances where you should and should not force the constant to be zero, but in many cases, such as the weight loss example, it is simply obvious.

3. **Why do you think that *y* is called the dependent variable and *x*'s the independent variables?**

For a regression equation such as $y = 0.121x_1 - 13.6x_2 + 0.020x_3 - 0.87$, y depends on the values of the x's, so y is called *dependent* (it is sometimes also called the *response* variable). I am not 100% sure why the x's are called *independent* (other than the fact that they are not the dependent variable) because they aren't really independent at all. For example, if you wanted to predict juvenile crime (y) in terms of education and parental income (x's), income and education are not independent of each other, they are strongly related. This is why I prefer the term *predictor* for the x variables: crime is the dependent variable; education and parental income are the predictors. You might also hear the term *covariate* used for the x variables. I tend to use covariate to mean "something we want in our regression but aren't really that interested in." As an example, imagine that we wanted to know whether occupational exposure to cigarette smoke (x) was associated with lung cancer (y). We'd obviously want to take into account smoking, because we know smokers have higher rates of lung cancer. But we aren't particularly interested in any estimate for the influence of smoking on cancer, because that has been reported many times before. So I'd call lung cancer the dependent variable, occupational tobacco exposure the predictor and smoking the covariate.

4. **In the marathon running example, the coefficient for "female" in the univariate analysis was 24, that is, women took 24 minutes longer to run the marathon. In the multivariable analysis, the coefficient was 23 minutes. How would you explain the difference between these two coefficients?**

In the univariate analysis, we are only looking at gender. The coefficient of 24 means: "Randomly select a man and a woman runner. Ignore anything else about them, such as their age and how far they run in training. Your best guess is that the man will run the marathon 24 minutes faster than the woman."

For the multivariable analysis, on the other hand, we are also taking into account age, and number of training miles. The coefficient of 23 therefore means: "If you had a man and a woman, and they were the same age and ran the same number of training miles each week, we would expect the man to run the marathon 23 minutes faster."

It is interesting to think through why the coefficients are different. It turns out that the women in the sample tended to run slightly fewer training miles than the men. The overall difference we saw between men and women of 24 minutes was therefore partly because women run more slowly than men, and partly because they didn't train as much. The coefficient from the multivariable analysis is the difference in marathon time "adjusted" for training, and gets closer to giving the natural difference in running speed between men and women.

5. *For enthusiastic students only:* **What is the log odds in logistic regression? What do statisticians tend to report instead coefficients for logistic regression?**

In a logistic regression, y is the log odds. In the text, for example, I used the equation for prostate cancer $y = 0.121\,x_1 - 13.6\,x_2 + 0.020\,x_3 - 0.87$ to calculate a y of -2.63 when x_1 was 4, x_2 was 0.25 and x_3 was 58. This means that the man in question has a log odds of cancer of -2.63.

To understand what this means we first have to understand odds. The probability of something is defined as the number of times it occurs divided by the total number of observations. If you do a study of 1000 children, of whom 750 pass an exam, you'd say that the probability of passing the exam was $750 \div 1000 = 75\%$. The odds of something is defined as the number of times it occurs divided by the number of times it doesn't occur. The odds of a child passing the exam are therefore $750 \div 250 = 3$. Here are some examples of probabilities and odds:

Probability of the event (e.g., passing the exam)	Probability of no event (e.g., failing the exam)	Odds of the event
50%	50%	1.0000
25%	75%	0.3333
10%	90%	0.1111
90%	10%	9.0000
5%	95%	0.0526
1%	99%	0.0101

One thing you might notice is that when something doesn't happen very often, the odds and the probability are similar. Something with a probability of 5% (0.05), for example, has an odds of 0.0526.

The formula relating probabilities and odds is that Odds = Probability $\div$ (1 − Probability). If you do the algebra, this means that Probability = Odds $\div$ (1 + Odds). But logistic regression gives you the log odds, that is, the logarithm (to base e) of the odds. So to get from the log odds to the probability, you first need to calculate the odds by raising e to the power of the log odds. In the example in the text, the log odds was -2.63; the odds are therefore $e^{-2.63} = 0.072$. This gives a probability of $0.072 \div (1 + 0.072) = 0.067$, or about 7%. To get $e^{-2.63}$ on most software, you type $\exp(-2.63)$. This means that if you have a log odds, and you want a probability, you type in something like $\exp(-2.63)/(1 + \exp(-2.63))$.

Coefficients for logistic regression, such as 0.121 for PSA, are a little hard to understand in isolation. So instead statisticians report what is called an odds ratio. To get an odds ratio, you raise e to the power of the coefficient. A coefficient 0.121 for PSA gives an odds ratio of $e^{0.121} = 1.13$. In other words, for every one point increase in PSA level, a man's odds of cancer increase by 1.13 (e.g. from 8% to 9%). This is often useful for giving a rough and ready idea of what drives risk.

Here is a quick example. Let's imagine that you were doing a study of job discrimination and you were examining the results of job interviews. Now assuming that all candidates were in fact well qualified to do the job, a regression equation might be something like $y = $ log odds of a job $= -0.223 \times$ African American $- 0.105 \times$ woman $-$ $0.020 \times$ Age in years $- 1$. This gives an odds ratio of $e^{-0.223} = 0.8$ for African American, $e^{-0.105} = 0.9$ for female and $e^{-0.020} = 0.98$ for each year in age. An odds ratio of 0.8 for African American means that, all else being equal, an African American has an odds 20% lower of getting the job (odds ratios less than one mean "less likely to get the job" and odds ratios of more than one mean "more likely to get the job"). So it looks as though African Americans, women and older people are indeed experiencing job discrimination in this experiment. It also looks as though discrimination is worse for African Americans than for women or older people. However, you have to be careful. Remember that the odds ratio for age is 0.98 *per year*. If you were one year older than another candidate, your odds would be 0.98, or 2% less; if you were two years older, the odds would be $0.98 \times 0.98 = 0.96$, or 4% lower; if you were 20 years older, your odds would be $0.98^{20} = 0.67$, or about a third lower. So it appears that, in this experiment, age was a bigger barrier to employment than gender or race.

CHAPTER 19
My assistant turns up for work with shorter hair: About regression and confounding

1. **In our multivariable regression, junk food was associated with obesity even after controlling for income, gender, education and exercise. Can we conclude that eating junk food causes obesity?**

It is very hard to deduce causality from statistical associations. The most obvious reason is one I discussed in the text: confounding. For example, have a look at the following diagram, which shows the stork population and birth rate in Germany in the early part of the 20th century:

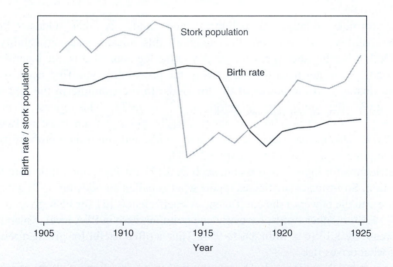

There is clearly a very strong relationship between the stork population and the birth rate. Indeed, the relationship is exactly what one would expect if babies were brought to mothers by storks, with the birth rate falling shortly after the stork population goes down. The confounder here is the First World War, which had an immediate impact on the stork population (due to bombs going off where storks normally fly around) and a delayed impact on the birth rate (men dying in the war cannot get their wives pregnant to give birth 9 months later).

The other reason why it is difficult to deduce cause from statistics alone has a nice analogy with romantic relationships. Mary and Craig are in a relationship—ok, but who caused it? Did Mary ask Craig or did Craig put the moves on Mary? Similarly, if there is a relationship between two variables, you can't deduce which caused which. In one set of data I analyzed, I found that an unhealthy diet was associated with lower rates of heart disease. This isn't because eating lots of fat and sugar is good for your heart. The most likely explanation is that patients with heart disease are told to watch their diet. If you meet a guy frantic about avoiding fatty foods, it might well be because he's been told that too much fat could kill him.

Incidentally, it is easy to get fooled by what you happen to call the predictor and the dependent variable. In the marathon running example (see *When to visit Chicago: About linear and logistic regression*), it is obvious that increased training causes runners to complete the marathon faster, not the other way around. However, the *p*-value for training miles in the regression y = marathon time = $b_1 \times$ training miles + $b_2 \times$ age + $b_3 \times$ gender is identical to the *p*-value for marathon time in y = training miles = $b_1 \times$ marathon time + $b_2 \times$ age + $b_3 \times$ gender. Just because you call something a predictor, it doesn't mean that it caused the dependent variable.

If you haven't done it already, just repeat the following phrase to yourself a few times: "Correlation does not imply causation." Do it carefully though. For many years the tobacco industry denied that cigarette smoking causes lung cancer on the grounds that studies showing increased lung cancer rates in smokers merely indicated a statistical association, and did not demonstrate cause and effect. The point about smoking is that there are multiple sources of data, not just statistical associations between smoking and cancer in humans. For example, chemical analysis of tobacco smoke shows that it contains known carcinogens, and exposing animals to tobacco smoke—when done carefully—leads to lung tumors. Moreover, there are numerous different types of human study and it is difficult to explain them all away. It is not unreasonable to postulate that people who choose to smoke engage in other behaviors that raise their risk of lung cancer, and it is these other behaviors that are to blame. But this would not explain why people who quit smoking have lower rates of lung cancer than those who continue to smoke, why people who smoke more are at higher risk, why lung cancer rates over time track smoking rates (e.g., they have started to fall as fewer people smoke) or why lung cancer rates in different countries vary by national rates of tobacco use. It is also hard to explain why smokers get cancers predominately in parts of the body that come into direct contact with smoke, such as the lips, tongue, throat and lungs.

Again, it isn't anything inherently statistical that does or does not allow us to draw a conclusion—such as whether or not smoking causes lung cancer—but a wider body of scientific knowledge. Statistics *helps* us draw scientific conclusions; it shouldn't *determine* our scientific conclusions.

2. **I gave diet and exercise as an example of something that couldn't be measured precisely. Couldn't we get people to complete a diary of everything they ate and all the exercise they did?**

    We could, but it wouldn't help much. The point is, you don't eat fattening foods and suddenly become obese. It is dietary and exercise habits over many years that lead to weight problems. We might be able to measure what someone is currently eating, but it is just about impossible to reconstruct a person's diet over a 20- or 30-year period. Similarly, we cannot ever measure accurately how much someone has exercised over their lifetime.

3. *For enthusiastic students only:* **I reported a coefficient of 0.334 and a standard error of 0.121. How did I get the odds ratio of 1.40? How did I get the confidence interval?**

    An odds ratio is simply *e* raised to the power of the coefficient: $e^{0.334} = 1.40$. The confidence interval for a coefficient is plus or minus 1.96, its standard error. This gives a confidence interval of 0.097 to 0.571; $e^{0.097} = 1.10$ and $e^{0.571} = 1.77$, so the confidence interval for the odds ratio is 1.10 to 1.77.

CHAPTER 20
I ignore my child's cough, my wife panics: About specificity and sensitivity

1. **In the experiment to determine whether my wife or I had a better approach to child health, we compared our results to that of a doctor. I described the results of this experiment in terms such as: "There was a total of 50 sick children; 49 of these were picked out by my wife." To be more accurate, I should have said: "A total of 50 children were described by the doctor as sick; 49 of these were picked out by my wife." In other words, I am using "described by the doctor as sick" to mean "really was sick." But is the doctor always right?**

Well, no, doctors are not always right (apart from my father-in-law, of course). The problem is that to work out sensitivity, we need to know who tests positive among patients who really do have the disease, and so we need some way to judge "really having the disease." In short, sensitivity and specificity are about comparing test results to the truth, but "truth" can be a bit of a slippery concept: without wishing to get overly philosophical, how do we ever know something is true for sure?

Statisticians get around this problem by using the term "gold standard." We choose something to compare against our diagnostic test, call this the "gold standard" and assume that this is the truth, although we know it might not be.

Here is an example from cancer research. Many cancer researchers are interested in whether high levels of particular proteins in the blood indicate prostate cancer. The best way to find out if you have prostate cancer is a biopsy, which involves inserting needles into the prostate, taking samples of prostate tissue and then looking at the tissue under the microscope to see if there are any cancer cells. In a typical study, we might look at the sensitivity and specificity of having high levels of a certain protein; we do so by comparing the results of the blood test with the results of the biopsy. It should be obvious that the biopsy is not 100% accurate. For example, the needles might miss the tumor and collect only healthy prostate cells. But the only way to find out for sure if a man has prostate cancer is to remove his prostate, cut it into sections and then examine the sections under the microscope (and no man wants his prostate removed if he doesn't have prostate cancer). So we just call the biopsy the "gold standard" as a reasonable approximation.

2. **Do sensitivity and specificity ever tell us which of two diagnostic tests is better?**

There are some cases where you can indeed use sensitivity and specificity to help you choose a diagnostic test. One obvious example would be if one test had both higher sensitivity and specificity than another. It is also easy to pick the best test if two tests have equal sensitivity, but one has better specificity:

Test	Sensitivity	Specificity
A	80%	40%
B	80%	50%

Test B will find just as many cases of disease as test A (equal sensitivity) but leads to fewer positive results in patients without the disease (better specificity), so test B is the better test.

Naturally, the converse is true. If the two tests had equal specificity but test A had better sensitivity, that is the one you'd choose.

The final example of where sensitivity and specificity can be useful is when the sensitivity and specificity of the two tests "mirror" each other, for example:

Test	Sensitivity	Specificity
A	85%	40%
B	40%	85%

All you have to do is decide whether sensitivity or specificity is more important and choose test A or B respectively.

That said, these sorts of situations—where it is easy to use sensitivity and specificity to choose the better of two diagnostic tests—are very much the exception. What is more common is where, say, sensitivity is more important than specificity and you get results like this:

Test	Sensitivity	Specificity
A	85%	40%
B	80%	50%

We'd probably want to go for test A, because we said that sensitivity is more important. But is a 10% loss in specificity worth a 5% gain in sensitivity? It can be difficult to say.

CHAPTER 21
Avoid the sales: Statistics to help make decisions

1. **Using a regression equation, Helen's doctor calculates that her risk of a heart attack is 8%. She is told that if she takes a cholesterol lowering drug, her risk will be reduced by 25%. However, the drug raises her risk of cancer by 0.5%. How much does Helen's risk of a heart attack decrease in absolute terms if she takes the drug? Do you think she should take it? What if her risk of heart disease was 2%?**

Thinking about relative and absolute risk is one time that a few formulas can be very helpful, especially as they are so simple.

Helen is told that her risk will be reduced by 25%. This is the relative risk *reduction*. To calculate her relative risk, we use:

$$\text{Relative risk} = 100\% - \text{Relative Risk Reduction}$$

This gives a relative risk of $100\% - 25\% = 75\%$. The definition of a relative risk is:

$$\text{Relative risk} = \frac{\text{Risk with the drug}}{\text{Risk without the drug}}$$

This means that:

$$\text{Risk with the drug} = \text{Risk without the drug} \times \text{Relative Risk}$$

We know that Helen's risk of a heart attack without the drug is 8%, so we can calculate her risk if she takes the drug as: $8\% \times 75\% = 6\%$. To calculate the absolute risk difference, we use:

$$\text{Absolute risk difference} = \text{Risk with the drug} - \text{Risk without the drug}$$

This gives $6\% - 8\% = -2\%$. Helen therefore has a greater decrease in her risk of heart attack (2%) than increase in the risk of cancer (0.5%). Unless Helen thinks that getting cancer is very much worse than having a heart attack (say, 3 or 4 times worse), she should take the drug.

That said, the benefit isn't very large. Even leaving aside the risk of cancer, for every 100 women like Helen, you can work out that 92 will be free of heart attack even if they refuse drug treatment and 6 will get a heart attack even if they take the drug. That leaves only 2 of 100 women who will benefit from treatment. In other words, on average, 50 women must take the drug in order for one to benefit. This is the *number needed to treat* and is calculated as $1 \div$ absolute risk difference.

If Helen's risk of heart attack was 2%, the risk with the drug would be $2\% \times 75\% = 1.5\%$, an absolute risk difference of $2\% - 1.5\% = 0.5\%$. In other words, the drug decreases Helen's risk of heart attack to exactly the same degree as it increases her risk of cancer. So it is hard to think why Helen would want to take the drug. Note that the relative risk reduction is the same (25%), giving another example of where relative risk (just like "30% off" at a sale) is of little help for decision making.

2. **I gave an example of decision analysis for a medical decision. However, decision analysis did not develop in medicine, but in another field of statistics. Which?**

   Decision analysis originated for helping business decisions. The theory is that, instead of just going with a gut reaction, a business person can identify all possible outcomes of different decisions, and then estimate the probability and amount of profit (or loss) of each. For example, for the decision tree shown in the text, the decision "bone marrow transplant" might be replaced by "market a new product," with "standard chemotherapy" replaced by "keep existing product line." Outcomes such as "major response" or "no response" might be replaced by "new product seen as a major improvement" and "no difference in sales," with profit figures calculated for each eventuality.

3. **Decision analysis isn't widely used. Why do you think not?**

   In order to complete a decision tree, you need to calculate probabilities and outcomes. In the case of bone marrow transplant, for example, we entered into the decision tree a 10% risk of death and a 60% chance of completing treatment. We also gave survival time of 10 months for a minor response and 60 months for a major response. I suggested that these numbers might come from "the scientific literature." One immediate problem is that these sorts of numbers might not be available or might be controversial. As an example from my own work, there is disagreement about just when a man should receive a biopsy for prostate cancer. We could create a decision tree to compare the decisions of "early" vs. "delayed" biopsy, but we would need to know the effects on a man's survival of a missed prostate cancer and the effects on his quality of life from an unnecessary biopsy. There is considerable disagreement between doctors on both of these points.

   The other reason why decision analysis is rarely used is interesting for what it tells us about how statistics works in practice. Decision analysis is considerably more difficult and complicated than the simple introduction I gave in the text might suggest (this is true of the book as a whole). As a result, it takes very specialized training to undertake decision analysis. This type of training is rarely included in traditional statistics courses, and as a result, many statisticians have very little knowledge of decision theory.

CHAPTER 22
One better than Tommy John: Four statistical errors that are totally trivial, but which matter a great deal

1. I mentioned that the sentence had "half an error." What was it?

The authors tell us that "baseline" age was no different between groups. This was a trial on pain, and in most pain trials, patients are on study for the same period of time. For example, patients fill in a baseline questionnaire, receive a bottle of pills or placebos to take daily and then fill in a second questionnaire about their pain six weeks later. So unless patients in different treatment groups grew old at different rates, there is no reason to tell us that it is "baseline" age that is being compared.

CHAPTER 23
Weed control for p-values: A single scientific question should be addressed by a single statistical test

1. What does our regression assume about the association between the dose of the drug and immune function?

Our regression equation was $y = 0.30x + 10.2$. If you plotted this out, you would get a straight line, so our assumption is that there is a linear association between dose and immunity. Here are the data that I used to create the regression:

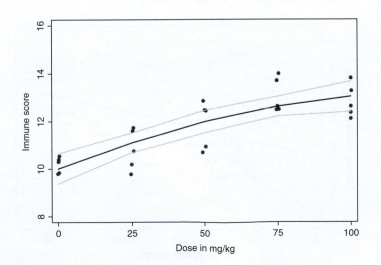

Dose in mg/kg

Looking at the graph, our assumption of a linear relationship between dose and immunity doesn't seem unreasonable. There are ways to check this other than by just "eyeballing." The easiest way to get a curve is to create a regression line of $y = ax^2 + bx + c$. When I do so, I get a -0.00018 for a, with a standard error of 0.000148. The coefficient isn't much bigger than the standard error, so the result is not statistically significant (the p-value is actually 0.2). Accordingly, we might say that we have no evidence for a non-linear association between dose and immune score if dose is 100 mg/kg or less.

One quick point about the graph: I added what is called *jitter*. Some of the mice had the same dose and very similar immune scores, so their points would have been one on top of the other on the graph. Jitter is a little bit of random noise that spreads out points so that they can be seen more easily.

2. **The authors of the immunity study found no significant difference between the 50 and 25 mg/kg doses. Does this really mean that "there is no difference between 50 and 25?"**

A *t* test comparing immune scores between the 50 and 25 mg/kg doses is testing the null hypothesis that the two doses have the same effect on immunity. A statistically significant p-value would lead us to reject this null hypothesis and conclude that 50 and 25 have different effects. A high p-value means that we are unable to reject the null; however, it doesn't mean that we can accept the null hypothesis (see *Michael Jordan won't accept the null hypothesis: How to interpret high p-values*). Being unable to prove a difference is not the same as proving no difference.

CHAPTER 24
How to shoot a TV episode: Avoiding statistical analyses that don't provide meaningful numbers

1. **I suggested that, for the crime data set, regression would give us a more meaningful number than correlation. Another alternative involves no numbers at all. How might you investigate these data without reporting specific numbers?**

You can also present data on a graph, which shows the data for each state, the regression line and the 95% confidence interval for the regression line.

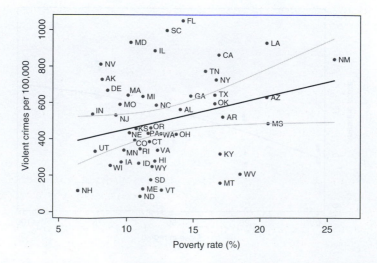

The nice thing about this graph is that not only does it show that the relationship between poverty and crime is not particularly strong, but it gives you an idea of where the action might well be. The states with the low crime rates—New Hampshire, North Dakota, Vermont, Maine and Montana—seem to be rural, whereas many of the high crime states—Florida, Illinois, Maryland and California—are more urban. If you run a regression on the proportion of the state's population living in a metropolitan area you get this:

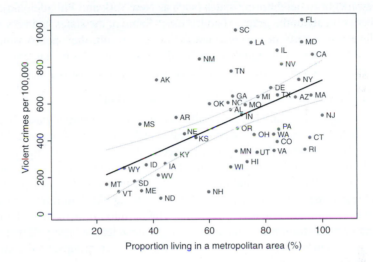

It certainly seems as though the proportion of patients living in metropolitan areas does a better job of explaining the data than the poverty rate. Some statisticians would recommend then calculating and comparing estimates of how well metropolitan living and poverty explain crime, using, for example, R^2 (see point 3, below). My own view is, why bother? Our aim isn't to have a dog show between different predictors to see which wins the crown of "Best Predictor," what we want to do is understand our dependent variable (y = violent crime) as best we can. So rather than providing individual correlation coefficients (or regressions) for each variable, one at a time (poverty, metropolitan living, etc.), we can put several variables into a single multi-variable regression.

Violent crimes per 100,000 =

22.2 × Percentage of state residents living in poverty

− 9.8 × Percentage of state residents with college education

+ 5.6 × Percentage of state residents living in a metropolitan area

− 3.2 × Median household income in $000's

+ 548

This multivariable regression gives a more complete picture of the factors that are associated with violent crime. It also helps deal with the issue of confounding (see *My assistant turns up for work with shorter hair: About regression and confounding*). For example, if you look only at crime and income it looks like crime rises with increasing median household income. However, there is a strong association between income and metropolitan living, with rural areas tending to be less wealthy. Once you adjust for metropolitan living, you see that increasing income is associated with lower crime rates, although not by very much.

Incidentally, "living in a metropolitan area" doesn't mean that you are in an apartment building downtown. The way that a "metropolitan statistical area" is defined by the US government is a city, or a county close to a city if that county has strong ties to the city (for example, if many people commute from that county to the city). Some of New Jersey is pretty rural, but all of New Jersey has significant links to cities in New Jersey (such as Trenton and Newark) or in neighboring states (such as New York and Philadelphia).

One other point about the graphs: I added *jitter*. Some of the states had very similar crime rates and levels of poverty or metropolitan living. As a result, their points would have been one on top of the other on the graph. As pointed out in answer 1 for chapter 23, *jitter* is random noise that spreads out points so that they can be seen more easily.

2. **Should you never use chi-squared, ANOVA or correlation?**

As a practicing statistician I have used all three techniques. I actually use chi squared quite a lot. (Truth be told, I tend to use something called Fisher's exact test rather than chi squared, but the differences between the two aren't particularly important.) I am just mindful that chi squared only gives a *p*-value and make sure that I find some way of giving an estimate as well. In the case of the religion and marriage data, for example, I would probably show the table, give the *p*-value from chi squared, and say something like, "To illustrate these findings, we categorized number of friends of the same religion into all or most vs. half or less than half, and categorized attitude as mind not at all vs. mind at least a little." I'd then give the estimates as described in the text (i.e., an 18% difference between groups; 95% confidence interval 13%, 22%).

I've used ANOVA sometimes in observational studies. For example, I've looked at a database put together by a surgeon who was keeping track of his results. He wanted to know if there were any differences in outcome between three alternative ways of doing the surgery. The first thing I did was to check whether there were any obvious differences between the sorts of patients receiving each type of surgery, something that might make it difficult to compare results. One of the things I checked was age and because I had a continuous dependent variable (age) and three groups (the three different types of surgery), I used an ANOVA.

I most often use correlation to examine relationships between different predictors. In one study, we were looking for proteins in the blood that might predict whether someone was at risk for cancer. It turned out that two of the proteins had an extremely high correlation (almost 0.95). As the level of one protein told you pretty much exactly what the level of the other protein was going to be, we concluded that it would never be worth it for a doctor to measure both.

Statisticians have developed a wide array of tools to help us understand data, and I'd be loath to describe any of them as being without any value at all. But, on the subject of tools, let me tell you that I own both a screwdriver, which I use a lot, and a crowbar, which I have used once or twice. If anyone told me that they used the crowbar a lot and the screwdriver rarely, I'd start looking around for a police officer. It isn't much different with statistics: we should spend most of our time using the tools that are most helpful for everyday tasks.

3. **Is correlation really a dimensionless number, like it taking "2.8" to set up a film shoot?**

A correlation does actually have some interpretations that are meaningful, although mainly to statisticians. The first concerns what is called "explained variation": the square of the correlation, R^2, tells you how much variation in one variable you can explain by looking at variation in the other. The square of 0.27 is 0.07, so the correlation of 0.27 between violent crime rates and poverty rates tells you that differences in poverty rates explain 7% of the differences in crime rates. This isn't a lot, as it means that about 93% of crime rates are unexplained. You can also work out an R^2 for a multivariable regression. The R^2 for the regression above (predicting violent crime rates in terms of income, poverty, metropolitan living and education) is about 50%.

Another way to interpret correlation is in terms of standard deviations. The correlation of 0.27 between poverty and crime means that a one standard deviation increase in the poverty rate leads to a 0.27 standard deviation increase in the crime rate (and the other way around). I think that this is somewhat difficult to understand. This is especially in comparison to regression, which gives you a result in real terms, such as that a 1% increase in the poverty rate is associated with 17 more violent crimes per 100,000. Talking of "17 more violent crimes" reminds us that we are dealing with real people who suffer the horrors of violence; a "standardized difference of 0.27" is a statistical abstraction that can lead us away from that reality.

4. Are we interested in inference for the crime data set? Should we report *p*-values from our analyses? What about confidence intervals?

We work out *p*-values by imagining what results we would get if the null hypothesis were true and we repeated the study a large number of times (see *The probability of a dry toothbrush: What is a p-value anyway?*). Similarly, the 95% confidence is defined as "if the study were repeated a larger number of times, 95% of the 95% confidence intervals would include the true estimate." All of this is on the assumption that when we repeat a study, the results may change. This is certainly true of an opinion poll (see *Statistical ties, and why you shouldn't wear one: More on confidence intervals*) or a study of hair length in men (see *Long hair: A standard error of the older male*) because we are calculating means and proportions from a sample. Once we select a different sample we might well get a different answer.

But in this case, our data set is the crime statistics for the different states in 1996. We aren't sampling 50 states from some imaginary larger population of states, we have all the data that there are, and if we repeated our study, we'd get exactly the same results. As a result, reporting *p*-values and confidence intervals doesn't make sense.

So, here is a rule of thumb: if you have the whole population, rather than a sample, don't report confidence intervals and *p*-values. One way to remember this is to think about Nathan's fourth grade class, which had 14 boys and 12 girls. When I plug these results into my computer, it tells me that 46% of Nathan's class are girls (which is true), but it also gives a 95% confidence interval of 27% to 67%. Computers are dumb like that—they spit out an answer even if the question is a stupid one: it isn't that our best guess is 46% girls, but it could be as high as 67% or as low as 27%; we know that the class is 46% girls, end of story. Also, over Christmas, one boy's family moves to the suburbs and leaves the school. My computer uses McNemar's test for paired samples to give a *p*-value of 0.8 for the null hypothesis of no difference in the proportion of girls comparing before and after Christmas. This suggests that we have failed to reject the null hypothesis and would have to tell the teacher that we have insufficient evidence of a change in gender ratio in the class.

The fact that we don't do so is an insight as to why we don't tend to think statistically in everyday life. When we say that prices at the Perch Cafe have gone up, or that the bridge was backed up with traffic, or that Peggy had a lot of people at her party, we aren't dealing with samples—we have all the data that we could get, that is, we have the whole population. Accordingly, we say these things with confidence and leave out the confidence interval.

CHAPTER 25
Sam, 93 years old, 700 pound Florida super-granddad: Two common errors in regression

1. **The regression for Sam's mile time was based on five data points, his time at ages 12, 14, 16, 18 and 20. Any thoughts as to whether this is a good or bad idea?**

   You'll see a lot written in textbooks about exactly how many variables you can put in a regression depending on the number of observations you have (for a linear regression) or events (for a logistic regression). One figure widely thrown around is that you need at least 10 data points or events for every predictor variable. For Sam's mile time we have 5 data points and 2 predictor variables (age and age^2), so we clearly have a problem. The name of this problem is *overfit*, which means that a regression will work very well for the current data set, but not for a new and different data set (e.g., if we looked up some old school records and found other mile times for Sam).

   I actually don't like to think of "overfit" and "events per variable" in overly statistical terms. To me it is simply a scientific rule of thumb: don't try to use too little to explain too much. As a simple example, imagine that you wanted to know whether food, atmosphere or service made the most difference to whether a restaurant was successful and used the restaurants in your home town as the data set. Except that there are only five restaurants:

Restaurant	Food	Atmosphere	Service	Successful?
Al di La	Excellent	Good	Ok	Very
Two Boots	Moderate	Good	Excellent	Very
Applewood	Excellent	Ok	Good	Quite
Song	Good	Excellent	Ok	Very
Cocotte	Good	Good	Good	No

   You might compare *Al di La* to *Cocotte* and conclude that it is the food that makes the difference; on the other hand *Two Boots* is successful despite so-so food. Comparing *Applewood* to *Song* suggests that atmosphere really counts, but that doesn't seem to explain why *Cocotte* closed down. In short, you just don't have enough data to inform three different hypotheses about what makes a restaurant successful.

2. **How did I work out that Sam could bench press a half a ton at age 93?**

   Sam could bench press 160 lbs at age 16 and 180 lbs at age 18. Assuming that a baby aged 0 cannot bench press any weight at all, this gives a simple regression equation of: Bench press (lbs) $= 10 \times$ Age (yrs). This would predict that Sam could bench press 930 lbs at age 93.

3. *For enthusiastic students only:* In the regression equation where y was time for the mile in seconds and x was age, I gave $y = 1.429x^2 - 54.2x + 784$. Why is the coefficient for $x^2 (1.429)$ given to three decimal places whereas only a single decimal place is given for the coefficient for $x (54.2)$?

The x is age, so x^2 is age^2 and can end up being very large. For example, for someone age 50, the coefficient for age^2 is multiplied by 2500. Reporting a co-efficient for x^2 of 1.43 rather than 1.429 makes a 2.5 second difference to our prediction. Reporting a coefficient for x of 54.2, rather than 54.21 (the coefficient rounded to 2 decimal places) makes only a 0.5 second difference to our prediction.

CHAPTER 26
Regression to the Mike: A statistical explanation of why an eligible friend of mine is still single

1. **Does regression to the mean explain why Mike is single?**

Actually, no, it doesn't—meaning that this chapter has been sold to you under rather false pretenses. What I explain using regression to the mean is why Mike's housemates all meet someone and move out. Most people in their 30's (the typical age of Mike's housemates) are in relationships and live with their partner. "Living with partner" is thus the "mean." Mike only rents his room out to singles, and thus to individuals who are not at the mean; a repeat observation of one of Mike's housemates (e.g., a year after starting to live in Mike's place) is likely to find that the individual has regressed to the mean, that is, they have met someone and moved in with them.

The real explanation for why Mike is still single is obviously that he hasn't met the right person yet. The other explanation is that, if he weren't single, he probably wouldn't be renting out his spare room and wouldn't particularly mind his housemates' romantic success even if he did. In other words, if Mike weren't single, I wouldn't be writing about him. If you are interested, this is a version of what cosmologists call the anthropic principle: we don't have to explain why conditions in the universe are just right for intelligent life because if they weren't, we wouldn't be around to wonder why not.

2. **What is the connection between regression to the mean and linear or logistic regression, the statistical technique used by statisticians to quantify relationships between variables?**

Linear regression and regression to the mean are quite separate statistical ideas and it would be natural to think that it is just coincidental that both include the word "regression." However, the two terms actually share the same historical origin. In the 19th century, the study of human heredity was known as *eugenics* because, roughly speaking, people believed that the human race could be improved by selective breeding. The British scientist who founded eugenics, Francis Galton, took measurements of families and then tried to work out the relationship between the heights of the parents and the heights of their adult children. He noticed that, although tall parents tended to have tall children, the children weren't quite as tall as the parents were; something similar was true of short parents. He called this "regression to mediocrity" (mediocrity being used as a term for "average"). But he also wrote out an equation to describe the relationship between the heights of parents and those of their children: y = difference of children's height from the mean = 2/3 difference of parent's height from the mean. For example, if a man's parents were 3 inches taller than average, he would generally be 2 inches taller than average. Because this equation described "regression to mediocrity," it was described as a "regression equation" and the term stuck. What hasn't stuck is eugenics, the hope of breeding a master race and the use of the word "mediocre" as average. Hence statisticians have replaced the term "regression to mediocrity" with "regression to the mean." "Regression to the Mike" is yet to gain widespread currency among academic statisticians, but maybe this will change.

CHAPTER 27
OJ Simpson, Sally Clark, George and me: About conditional probability

1. **I described two events as independent if information about one gives you no information about the other. Similarly, two variables are independent if information about one gives you no information about the other. What is the relationship between independence and statistical tests such as the *t*-test or chi-squared?**

    All of the common statistical tests assume that the data are independent. Applying these tests to non-independent data is a very common error. An obvious example is repeat observations. For example, imagine that we were interested in the influence of shop design on buying habits, and compared the sales at two branches of the same clothing store. We might get the following data.

	Atlantic Terminal: "Open" entrance area	Broadway: Entrance area with clothing racks
Monday	$2,800	$4,400
Tuesday	$2,600	$4,300
Wednesday	$3,200	$4,700
Thursday	$3,000	$4,500
Friday	$3,900	$5,700
Saturday	$6,000	$7,200
Sunday	$5,400	$7,000

You might be tempted to compare these sales data by a *t* test or Wilcoxon (you get *p* of around 0.045 either way). But this is to forget that a store's sales on one day are not independent of its sales on another—if I tell you Monday's sales figures, you can take a reasonable guess at Tuesday's. The easiest way to think about why a *t* test is not appropriate here is to stop thinking about statistics altogether. What would you say if you were the CEO of the clothing company and one of your managers brought the data to you claiming that putting clothing racks in the entrance area was a good idea? You'd probably say something like, "This isn't much data at all. You have only shown me two stores and it doesn't surprise me that the store in mid-town Manhattan does better than the one in that second rate mall in Brooklyn." Translating this into statistics speak: you don't have 14 independent data points, you have two sets of seven non-independent data points.

Another example of non-independence is where something can affect multiple individuals. Take the case of study to determine whether memorizing multiplication tables affects math scores in elementary school. One approach might be to compare the results of two classes, one taught by a teacher who encouraged children to remember the multiplication tables and the other who did not. If there were 25 students in each class, you would end up with a total of 50 scores. However, again, the data are not independent. If I tell you that one kid in a class did very well, you might guess that the teacher was a good one and that other kids in that class did well too. As a result, you can't just enter the 50 numbers into a *t* test or Wilcoxon.

What this all comes down to, of course, is experimental design. It is pretty silly to design a study on store layout so that it includes only two stores because a store's sales depend on all sorts of things other than whether the entrance is left open or includes clothing racks. Similarly, teachers differ in all sorts of ways so it will be difficult to attribute any differences in the results of two teachers to a single difference in the use of a specific teaching technique.

2. **In the text I describe Professor Meadows' argument as "only about 1 in 8500 babies die of crib death . . . the chance of two crib deaths [is] 1 in 8500 multiplied by 1 in 8500 [or] 1 in 73 million." I pointed out that the "1 in 73 million" number (a) assumes that the chance that an infant will die of crib death is independent of the chance that a sibling would; and (b) says very little about the probability that the children were murdered. There is an even more fundamental mathematical problem though. Any thoughts?**

You might have noticed- 8500 × 8500 is nearer to 72 million, not 73 million. I'd love to be able to say about Meadows, "Can't do statistics? He can't even do basic math!" but I don't have a transcript of the court case so I don't know exactly what he actually said. It is interesting nonetheless that all of the very many newspaper stories reporting the Sally Clark case repeated a basic multiplication error.

3. **The Sally Clark case has been described as an example of the "Prosecutor's Fallacy." What do you think this is?**

The Prosecutor's Fallacy goes like this:

- Based on the evidence I have provided, it is highly unlikely that Mr. Shapiro is innocent.
- Therefore, Mr. Shapiro is guilty.

The problem with this argument is that there are two possibilities: Mr. Shapiro is innocent and Mr. Shapiro is guilty. We can't look at the probability of just one of these and draw conclusions about the other. Instead, we need to compare the probability of each.

For example, imagine that a robbery is committed and some DNA is collected at the scene which matches that of Mr. Shapiro. However, it turns out that at the time of the robbery, Mr. Shapiro was in a city thousands of miles away giving a presentation to the National Association of Police Officers. It is pretty unlikely that Mr. Shapiro's DNA would match that of a sample collected from a crime scene. The Prosecutor's Fallacy would be to conclude that Mr. Shapiro is guilty on those grounds alone. But it is even more unlikely that Mr. Shapiro could have committed the crime given that he had the world's most perfect alibi. So out of the two possibilities, we choose the most probable and declare Mr. Shapiro innocent.

The Prosecutor's Fallacy is often found in conspiracy theories. Take the idea that the works of Shakespeare were written by someone other than William Shakespeare. The argument against Shakespeare being Shakespeare is that he was relatively low class and it seems improbable that someone of his background could have written such worldly and brilliant plays. Accordingly, his plays must have been written by someone of higher social status, say, the seventeenth Earl of Oxford. Which is fine until you realize that the Earl of Oxford died before some of Shakespeare's plays were written. It seems more likely that Shakespeare was indeed Shakespeare than that the Earl of Oxford wrote plays while dead (how exactly would he get ink and paper in his coffin anyway?).

This is a great example of the Prosecutor's Fallacy:

- It is improbable that William Shakespeare could have written all those fantastic plays.
- Therefore William Shakespeare is not the author of Shakespeare's plays.

The counter-argument is to note that there are various competing theories and to see which one makes more sense:

- It is improbable that William Shakespeare could have written all those fantastic plays, that is certainly true.
- It is impossible for the Earl of Oxford to have written plays while dead.
- Therefore it is more likely that Shakespeare's plays were written by Shakespeare than by the Earl of Oxford.

Shakespeare himself provides a great example of the Prosecutor's Fallacy. In the play *Othello*, the lead character is a dark-skinned "Moor" who marries a white woman ("the fair Desdemona"). Desdemona's father, Brabantio, can't believe that his daughter would choose to marry a black man of her own free will and accuses Othello of witchcraft. He goes before the senators of Venice and says:

> *She is abused, stol'n from me, and corrupted*
> *By spells and medicines bought of mountebanks;*
> *For nature so preposterously to err,*
> *Being not deficient, blind, or lame of sense,*
> *Sans witchcraft could not.*

Loosely translated, this means that it is very unlikely that a white woman who wasn't blind, stupid or completely devoid of sense would get together with a black guy (an error against nature). Therefore it must be that Othello used magic and gave Desdemona potions bought from quack doctors. This is pretty close to the Prosecutor's Fallacy: one thing is unlikely (Desdemona actually liking Othello), so something else (Othello using witchcraft) must be true.

CHAPTER 28
Boy meets girl, girl rejects boy, boy starts multiple testing

1. Is subgroup analysis a problem mainly for medical research?

I used a medical example in the text because doctors seem particularly obsessed by subgroups. But subgroup analysis crops up in many fields of statistics. In sociology, researchers might be interested not only in the overall effect of a change in policing on overall crime, but on its effect on different types of crime; an educational psychologist might be interested in the effect of a teaching technique separately for boys and girls; a biologist might divide results by subspecies.

One of the most difficult areas for multiple testing is genetics. Everybody has thousands and thousands of genes, so it is likely that many of them will end up correlating with traits like, say, height, intelligence or athletic ability. A typical genetic study might examine 15,000 genes and we'd expect that 5%, around 750, would have a statistically significant association with the trait we were investigating, just by chance. Genetics researchers need to use special statistical analyses to deal with this problem, but they don't always do so properly.

However, the award for most gratuitous use of subgroups goes to ... sports. Batting averages tell you something I guess, and I suppose it makes sense to look at averages separately for infielders and outfielders. But do we really need to know which infielder has had the highest batting average since the all-star break? And no doubt drawing a walk is an important part of baseball, but should we care who has worked out the most 9th inning walks? My favorite example was in an NFL divisional play-off game, when it was reported that the visiting team had lost their previous eight games in the Eastern time zone. They won in a rout.

2. **Do all clinical trials have equal numbers of patients in each group (e.g., 1000 patients on the new drug and 1000 patients in the control group)? Do clinical trials include exactly equal numbers of men and women?**

No, research with humans almost never works out so cleanly. It is even difficult to get exactly the right number of patients in a study, let alone the same number in each group. For example, imagine I was running a study that aimed to get 2000 patients on study and that, at 11 am on the first Thursday in February, the 2000th patient was finally recruited. I am in a meeting at the time and don't get the message until 3 pm, when I send out an e-mail to all the doctors in the trial telling them to stop entering patients. Except that at 3 pm, doctors aren't reading their e-mails—they are sitting in clinic and trying to get patients to join important new medical studies. So by the time the doctors read their e-mails, they have probably entered at least a few more patients, leaving us with more than 2000. This is especially true of a certain Dr. Jones, who has changed his e-mail address and doesn't get the message until the following Wednesday. Then it turns out that some patients in the study were actually not eligible and that some died or moved before recurrence could be assessed.

All of which is to say that, although they make statistics much easier to understand, don't expect nice round numbers when you actually start doing research.

3. *For enthusiastic students only:* **One table in this chapter shows the probability that at least one of a given number of tests will be statistically significant if the null hypothesis is true. I worked out these numbers using the formula $1 - 0.95^n$ where n is the number of tests. This is the same as saying, "What is the probability that all tests have p-values ≥ 0.05?" (which is 0.95^n) and then saying, "The chance that at least one test has $p < 0.05$ is 1 minus that probability, i.e., $1 - 0.95^n$." As I hinted in the text, this formula is an oversimplification. Why?**

It is common to calculate the probability of multiple events (such as throwing two coins and getting heads both times) by multiplying the probability of each event separately (i.e., 0.5×0.5). This only gives the correct probability if the different events are independent, that is, if you can't guess the likelihood of one event given knowledge of the other. This is certainly true of coin tossing. If I tell you I just threw a heads it doesn't help you predict the next throw.

Different types of subgroups are not independent. If we do one subgroup analysis by age (old vs. young) and another by gender (male vs. female) it is the same people in both analyses, just divided up in different ways. As such, the results of subgroup analyses are no longer independent. As an example, imagine that I told you that a drug was ineffective but that in the clinical trial, just by chance, men on the drug had statistically better results than men on the placebo. If I asked you for the chance of a statistically significant difference between groups among the women, your reasoning should go:

- We know that the drug is ineffective, therefore the null hypothesis ("no differences between groups in the trial") is true.
- The probability of a statistically significant difference between drug and placebo groups if the null hypothesis is true is 0.05.
- No men are women, so the results of the subgroup analysis on men are irrelevant.
- Therefore the probability of a statistically significant difference between drug and placebo in a subgroup analysis of the women is 0.05.

However, what if I asked you about the results of a subgroup analysis for older people?

- We know that the drug is ineffective, therefore the null hypothesis ("no differences between groups in the trial") is true.
- The probability of a statistically significant difference between the drug and placebo groups if the null hypothesis is true is 0.05.
- The probability of a statistically significant difference between drug and placebo in a subgroup analysis of older people should be 0.05.
- However, some of the older people are men and we know that men on the drug did better than men on placebo.
- Therefore, the probability of a statistically significant difference between drug and placebo in the subgroup analysis of older people will be a little bit higher than 0.05.

In other words, the results of the subgroup analysis "what is the effect of the drug on older people?" are not independent of the results of the subgroup analysis "what is the effect of the drug on men?" To get a probability that at least one of the two analyses would be statistically significant (or both significant, or neither significant), you can't just multiply probabilities together; you have to do something more complicated (which I won't go into).

CHAPTER 29
Some things that have never happened to me: Why you shouldn't compare p-values

1. **Why might a stronger effect lead to a higher *p*-value, and less evidence against a null hypothesis of no effect? How might you explain the differences in *p*-values for the four examples in the text (taxis, movies, engine parts and blood pressure drugs)?**

We might expect that a stronger effect would lead to a lower *p*-value and, indeed, this is generally what you see. Generally, but not always—to think why not, we have to consider all the things that can affect a *p*-value.

a. **Sample size.** Larger studies provide more evidence. If the null hypothesis is false, larger studies will tend to have lower p-values than smaller studies.

b. **Standard deviation.** More variation means less confidence—that is, a wider confidence interval. Higher standard deviations lead to higher p-values.

c. **Study design.** Different studies can have different results due to differences in design (or to put it more simply, flawed studies can come to the wrong conclusion).

d. **Chance.** The p-value will vary every time you repeat a study and it is quite possible that, by chance, a stronger effect will be associated with a higher p-value. Indeed, it is quite possible in the blood pressure trial that a completely ineffective drug could have a statistically significant p-value compared to placebo, with a very effective drug being non-significant.

We can now apply each of these to our four examples of comparing p-values:

a. **Sample size.** There are more moving parts in most engines than parts that don't move. More parts mean a larger sample, which means more evidence and a lower p-value.

b. **Standard deviation.** In the blood pressure example, new drug A reduced blood pressure fairly predictably in just about everyone (mean reduction 6.5, standard deviation 0.5). Response to new drug B was much larger, although a lot more varied (mean reduction in blood pressure 15, standard deviation 12.5). Because the standard deviation for new drug A was so much lower than for new drug B, the p-value for the null hypothesis "new drug no more effective than standard treatment" is smaller.

c. **Study design.** In the hailing cabs example, the author was comparing two completely different studies (one by Bloggs in Chicago and the other by Smith in New York). There are all sorts of reasons why differences between the studies may have led to different results including the time of day, the part of town, how the participants were dressed, their age, their physical size, the time of year and so on. The fact that one study was conducted in New York and the other in Chicago is just one of many possible points of difference between the two studies.

d. **Chance.** In the study on movies, the p-value for the null hypothesis that "attitude to violence in movies changes with age" was 0.002 for women and 0.005 for men. These p-values are actually very close. Just for illustration, if there were 200 men and 200 women, half of whom were older and half younger, a difference in p-values between 0.002 and 0.005 could be caused by a single person changing their answer. So it is a bit of a stretch to interpret this small difference as indicating an effect that is "more pronounced in women than men."

2. **How would you test whether one effect was stronger than another? For example, how would you test whether women's attitude to violence in movies changes more with age than men's?**

One of the key points of the chapter is that if you want to test a hypothesis, you need a single p-value rather than comparing two different p-values. To test a hypothesis as to whether one effect is stronger than another, what you use is called a *test for interaction*. If you think that one predictor (e.g., gender) makes a difference in the effects of another (e.g., age), then the interaction term is when the two predictors are multiplied together. For example, you might have a data set that looks like this:

Female (1 if female, 0 if male)	Age	Interaction term	Is there too much violence in movies? (1 if yes, 0 if no)
1	62	62	1
0	34	0	0
1	25	25	0
etc.			

The interaction term is Female × Age, which ends up being just age for women and zero for men. The next step is a logistic regression y = attitude to violence in movies = b_1 × Age + b_2 × Female + b_3 × Interaction term. The p-value for b_3, the coefficient for the interaction term, tests whether attitude to violence in movies changes more with age in women than men. The p-value for the interaction term in my data set is 0.7, providing little evidence against the null hypothesis.

Incidentally, a graph is a nice way of illustrating this sort of data.

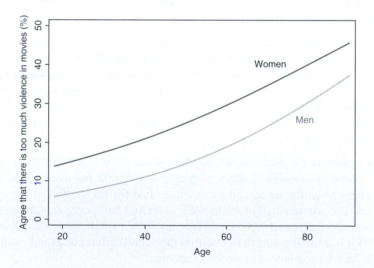

Clearly women tend to have a more negative attitude to violence in movies than men; it is also obvious that attitudes change with age. But it is not at all clear that women's attitudes change with age any differently than men's do—the lines look pretty parallel.

Here is a classic example of interaction to help you think through what it means. Some new cancer drugs work by "locking on" to a receptor on the surface of the cancer cell. As it happens though, some cancers do not have the receptor. Take the following data set from a study of the drug in mice. In the study, mice have cancer cells injected. Half are then treated with the drug and the size of the tumor is measured after 28 days.

Number of mice	Received drug?	Cancer cells had receptor?	Mean tumor size
4	Yes	No	20.2 mm
4	Yes	Yes	7.0 mm
4	No	No	21.2 mm
4	No	Yes	19.7 mm

The drug doesn't seem to work for cancers without the receptor, and the receptor on its own seems to make no difference to tumor size. The only time we see a reduction in cancer is when the drug is given and the cancer has the receptor. One analysis might be to look at the effect of the drug separately in cancers with and without the receptor: I get p-values of 0.02 and 0.6 respectively. But this is two p-values to test a single null hypothesis "the effects of the drug do not depend on the receptor." The alternative is a linear regression of y = tumor size = $b_1 \times$ Drug + $b_2 \times$ Receptor + $b_3 \times$ Interaction term, where the Interaction term is Drug $\times$ Receptor (i.e., 1 if mouse had receptor and was given the drug and 0 otherwise). If I run this on the data set, I get coefficients b_1 and b_2 as non-significant and b_3 as statistically significant, suggesting that the drug will likely only work in patients who have cancers with the receptor.

CHAPTER 30
How to win the marathon: Avoiding errors when measuring things that happen over time

1. **How do you think you should analyze the data from the job satisfaction study?**

   The correct analysis has an odd name—*landmark analysis*—but is very simple in principle. As you might guess, you just start the clock at a set time (the landmark). Employees complete their questionnaire after three months on the job, so that is when you start the clock. You exclude from your analysis anyone who quits before three months and also anyone who fills in their questionnaire late.

   Here is what the graph looks like:

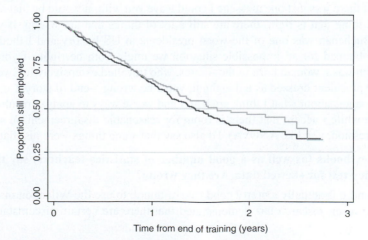

The whole graph is shifted to the left because we are now measuring from the end of training, not the date of recruitment. You can also see that the two curves are much closer together and there is now no statistically significant difference between groups.

2. How do you deal with lead time bias?

Like so many other scientific problems, the lead time bias problem is best fixed with a control group. If people who get screened are less likely to die of cancer than those who don't get screened, then you'd have pretty good evidence that screening was helpful. In the case of lung cancer screening, another cancer researcher (who happens to be a colleague of mine) has shown that people who get screened probably don't live any longer than people who don't get screened.

Incidentally, it also turned out that the researchers who published the study on lung cancer screening had patents on devices to find lung tumors (which they didn't tell anyone about), used tobacco industry money to fund their trial (which they didn't tell anyone about) and weren't entirely candid when reporting who died of what. Now I generally take the position "never attribute to conspiracy what you can attribute to a simple screw-up." Nonetheless, when you see bad statistics, it is worth wondering who stands to gain.

CHAPTER 31
The difference between bad statistics and a bacon sandwich: Are there "rules" in statistics?

1. So there is no right and wrong in statistics? Does that mean that anything goes and you'll get 100% on your exam even if you do a dumb analysis?

I guess if you were a true scientist you could do an experiment on your exam and see what happens, but I wouldn't suggest it. I could get all philosophical and discuss absolute and relative theories of truth (for example, you could say that even if reality is relative, it is still reality). Or I could point out that the opposite of "true for all people at all times" isn't "anything goes." It is perfectly reasonable for groups of people to ask members of their group to act in certain ways and to rebuke them if they don't. (A good example is dueling. This used to be legal and now it isn't, showing that ideas of right and wrong can change over time. However, that doesn't make it ok to start firing pistols at 20 paces.)

But if there was just one message I could leave you with, it would be that even if there isn't just one thing that is right, there are still a lot of things that are wrong. If you thought that James Buchanan was one of the worst presidents in US history and I thought that he was unfairly blamed for an impossible situation we might both be right; if you thought James Buchanan was a woman born in the Azores, who travelled extensively in Sweden before running for president dressed as a doughnut, you'd be wrong—end of story. Comparably, I have written papers about what I think are better and worse ways to analyze results from a clinical trial. Yet while I accept that there is room for reasonable disagreement on some methods (I like regression, you like ANOVA), I'd also say that some things were just flat out wrong.

2. Many textbooks (as well as a good number of statistics teachers) say that you should avoid the _t_ test for skewed data. Are they wrong?

No, and it is actually a pretty good rule of thumb to use the Wilcoxon instead for skewed data. However, research has demonstrated that there are certain circumstances where data are skewed and yet it is perfectly fine to use the *t* test. This makes two important points: (a) statistics is subject to research, not a set of rules and (b) things are often quite a bit more complicated than introductory textbooks make out. They are also more interesting, which is one reason why you might consider taking additional statistical classes.

3. **In the text I said that there aren't rules, laws and commandments about statistics, you just have to know what the latest statistical research shows. Does this suggest that all scientists need to look up the statistical journals before running a simple statistical analysis?**

No, but I'd expect that they'd use some textbooks or web-based materials to guide their analysis and I'd hope that these were based on the latest evidence. This might not be true for scientists graduating shortly after Pearl Harbor, who were still using their college textbook.

Actually, it might not even be true if their college textbook was published more than a few years ago. Here is a true story. A few years into my career, I looked up how to do an analysis in my college textbook. This had been published four years before I'd started graduate study and, taking into account the time I'd been in college, the text was about a decade old. It was completely obvious to me that the recommended analysis was inefficient, so I wrote to the author. The author agreed that things had moved on and we agreed to write a review article describing more up-to-date techniques. This article has been widely cited by other scientists.

Either way, this chapter is not a prescription for doing one thing or another as regards conducting a statistical analysis. It is more about being aware that statistics is not a set of hard and fast rules set in stone, but a science that changes over time.

CHAPTER 32
Look at your garbage bin: It may be the only thing you need to know about statistics

1. ***For enthusiastic students only:*** **Is there anything I need to know about statistical programming, other than the fact that I shouldn't bring this up on a first date?**

Probably not, unless you would ever consider doing a statistical analysis for some important purpose, like a thesis, scientific paper or business report.

Pretty much everyone running a statistical analysis today uses a computer to do so. To run an analysis on professional statistical software you have one of three choices:

   a. Type in a command such as "ttest handwriting_score, by(extra_teaching)."

   b. Use a pull-down menu to find "ttest," and then select "handwriting_score" and "extra_teaching" from the list of variables in the dialog box.

   c. Type one or more commands into a separate programming file and then run the programming file.

Most professional statisticians pretty much never use option (a) or (b), and always program every analysis. Just as a quick example, here is some code written for the software package STATA. You don't have to understand the details, just try get the big picture:

```
1.    clear
2.    *load up the data set
3.    use "handwriting data.dta", clear
4.    *exclude children from this analysis if they recently joined the school
5.    *this exclusion is described on page 4 of the study protocol
6.    *"Recent" means less than 3 months
7.    drop if time_at_school < 3
8.    *children with missing data assumed not to have received extra teaching
9.    *"extra_teaching==." means "no data for the extra teaching variable"
10.   replace extra_teaching = 0 if extra_teaching ==.
11.   *now run the ttest
12.   ttest handwriting_score, by(extra_teaching)
13.   *print out the results, rounding to a single decimal place
14.   "r()" means result. E.g. "r(se)" means, give the standard error
15.    display "The difference between groups is: " round(r(mean1) – r(mean2),.1)
16.    display "The 95% confidence interval is ",
17.    display round(r(mean1) – r(mean2)-r(se)*1.96),.1),
18.    display " to ",
19.    display round(r(mean1) – r(mean2)+r(se)*1.96),.1)
```

Some things to notice. The *'s on lines 2, 4, 5, 6, and so on indicate a comment. The computer ignores these lines, but they are very useful for programmers to keep track of what they are doing and why. This bit of code is pretty typical because as about half the lines of code are comments. In fact, pretty much every line of actual code is carefully explained. Also notice that the code includes a command (line 3) that opens up the data file. This is important because you want to run your analyses on the correct data set.

Another key point is that lines 15–19 print out the results of the t test.

Typical printout from statistical software for a t test looks like this:

	Obs	Mean	Std. Err.	Std. Dev.	[95% Conf. Interval]	
Group1	30	2.21	.1971801	1.08	1.806721	2.613279
Group2	30	1.566	.1811136	.992	1.195581	1.936419
combined	60	1.888	.1391911	1.078169	1.609479	2.166521
Diff		.644	.2677352		.1080698	1.17993

```
diff = mean(Group1) – mean(Group2)                              t = 2.4054
Ho: diff = 0                                        degrees of freedom = 58
Ha: diff < 0              Ha: diff != 0                      Ha: diff > 0
Pr(T < t) = 0.9903        Pr(T > t) = 0.0194        Pr(T > t) = 0.0097
```

This is rather confusing, and you see how it would be easy to make some mistakes cutting and pasting this into your word processor. The programming code, however, would print out something that would look like this:

The difference between groups is 0.6. The 95% confidence interval is 0.1 to 1.2.

The overarching issue here is that of reproducibility. To reproduce your analysis, you'd simply load up the code and press "go," and get exactly the same results.

One final thing: writing code seems time consuming, and it is. In the long run, however, it saves a lot of time. For example, imagine that the code was used in a research study and the authors sent the results to a scientific journal for publication. The journal editors write back to say that they like the study, but would be interested to know whether the results are "sensitive" to the exclusion of children new to the school. In other words, do the results change if all children are included, or if the criterion for "new" is changed from three months to one month? To find out, all you'd have to do is change line 7 of the code and run it again. This takes no time at all and will be 100% accurate. In comparison, without code you'd have to manually load up the data set manually, type a command to drop children who joined the school within a month of the experiment, type another command to define children with missing data as not having received extra teaching, select *t* test from the pull-down menu and select the right tick boxes and finally cut and paste the right bits of the printout. Not only is this very time consuming, but it is easy to make a mistake.

CHAPTER 33
Numbers that mean something: Linking math and science

1. **Jonas goes to the market and buys a 10 lb watermelon and 9 apples weighing 2¹/₂ lbs. He calculates the mean weight of the fruit as 12¹/₂ ÷ 10 = 1¹/₄ lbs. What are your thoughts about this statistic?**

   One reason we calculate statistics such as the mean is because they help us understand something. For example, if I said that the mean age of children taking part in a reading study was 13 with a standard deviation of 0.5, it would instantly give you an idea that this was a middle school project. The other reason to calculate the mean is because it helps us decide something, such as the best route home, how much to set aside for next year's budget or whether a painkiller is strong enough to make it worth taking.

   It is unclear what "mean fruit weight 1¹/₄ lbs" refers to, or how it could be used. Does a mean of 1¹/₄ lbs give us any idea of the sort of fruit that Jonas bought? None of the fruit weighs anywhere near 1¹/₄ lbs and if Jonas had reported the standard deviation (which was 3), we wouldn't be able to say that "95% of the fruit weighs within 2 standard deviations of the mean" because this would mean that some fruit weighed less than nothing at all. We also couldn't use the mean for any helpful purpose. For example, the mean doesn't answer "Jonas is going to buy 20 pieces of fruit; how much will his bag weigh?" because if he buys 20 pieces of fruit, he'll probably still only get 1 watermelon. In short, the mean weight of Jonas's fruit is little more than an abstract mathematical calculation, which has little use or meaning in the real world.

   You might think, "Oh, that's just stupid. No one would try to average out the weight of a watermelon and an apple." But actually, this sort of thing happens all the time. As an example, two doctors at my hospital reviewed all the scientific papers that had ever been published on recovery after a particular sort of cancer surgery. They worked out the recovery rate reported in each study and then took the average. Now working out the "average" recovery rate turned out to be very complicated statistically and a special technique called *meta-analysis* had to be used. After months of hard work, I was invited to see the results and comment on all the esoteric statistics. I didn't even look at the statistics—"recovery" had been defined totally differently in pretty much every scientific paper, so combining their results, and taking an average, was little more than weighing a bag containing some apples and a watermelon and calculating the mean weight of fruit.

2. **The surgeon concluded that "obesity may have some effect on survival." Words like "may," "might" and "could" are often found in the conclusion of scientific studies. Why should scientists avoid using these words?**

Think of some study—doesn't matter what—and come up with a possible conclusion. Here is one I just thought of: "Students learn more statistics from reading *What is a p-value anyway?* than from any competing statistics textbook." Now take your conclusion and add in a word like "may," "might" or "could" and see what you get. My conclusion becomes "Students *may* learn more statistics from reading *What is a p-value anyway?* than from any competing statistics textbook." What is interesting about this conclusion—and I am sure you'll find the same with your own example—is that it seems to be saying something provocative (that is, my book is fabulous) but in fact says nothing at all. The statement will be true in all cases except in the unlikely circumstance that we had proved it false. For example, no one has done a study to demonstrate that *People* magazine is not a good source for statistical knowledge so we can be confident in claiming that "students may learn more statistics from reading *People* magazine than from reading a statistics textbook." On which point, it is also true that "students may learn more statistics from sitting on the couch playing computer games than from turning up for class."

Words like "may" are actually vague in two very different ways. For example, if I say that "stretching the thigh muscles may increase sprinting speed," I might mean "we don't know for sure whether or not stretching leads to faster sprints," or I might mean "it is definitely true that stretching causes improvements in sprint times, but it doesn't help everyone, so we don't know for sure whether it will help you."

What is particularly dumb about using words like "may" or "might" in the conclusion of a scientific paper is that doing so suggests that the study was a total waste of time. Imagine that we analyzed data on thousands of car crashes and concluded that "cell phone use may increase the risk of a road traffic fatality." Well, yes, we knew that before we started—if yapping on a cell phone couldn't possibly increase the risk of driving into a tree, we wouldn't have done the study in the first place. This is particularly painful in my own field of cancer research. I have sometimes seen reports of large clinical trials concluding that "a second round of chemotherapy might improve cancer survival." Of course more chemotherapy "might" improve cancer survival, that much is totally obvious—chemotherapy kills cancer cells. So you put hundreds of patients through the drudgery of additional chemotherapy to find out something we already knew?

The reason why words like "may," "might" and "could" are so popular is that it absolves the author from any responsibility whatsoever. If I conclude that "a second round of chemotherapy may improve cancer survival" and someone else does a much better study proving that it doesn't, I can always say, "Hey, I never said that extra chemotherapy worked for sure, I only said that it might." This is the sort of thing that scam artists live by. A claim that "this simple real estate trick that could make you thousands of dollars richer" allows me to take your money, teach you nothing of value, and then defend myself with, "I didn't say that my program *would* make you money, only that it *could*."

We surely want science to be different from get-rich-quick schemes. If so, scientists need to stand by their conclusions and avoid saying only that something may, might or could be true.

Credits and References

Chapter 6

Page 23, Bar graph: results of a clinical trial of acupuncture for headache. Vickers, Andrew et al., (2004, March). "Acupuncture for chronic headache in primary care: large, pragmatic, randomized trial." *British Medical Journal, 328*(7442): 744. Reproduced with permission from BMJ Publishing Group Ltd.

Page 24, Graph: the reduction of headache during acupuncture. Trial and discussion of data on the same page. Vickers, Andrew et al, (2004, March). "Acupuncture for chronic headache in primary care: large, pragmatic, randomized trial." *British Medical Journal, 328* (7442): 744. Reproduced with permission from BMJ Publishing Group Ltd.

Chapter 7

Pages 26-30, "Why Does Chutes and Ladders Explain Hemoglobin Levels? Some Thoughts on the Normal Distribution." *Medscape*; WebMD, 111 8th Avenue, New York, NY 10011.

Page 28 (bottom), Graph: distribution of hemoglobin in a cohort of Swedish men. Lilja H., Ulmert D., Björk T., Becker C., Serio A.M., Nilsson J.A., Abrahamsson P.A., Vickers A.J., Berglund G. Long-term prediction of prostate cancer up to 25 years before diagnosis of prostate cancer using prostate kallikreins measured at age 44 to 50 years. *Journal of Clinical Oncology*, 2007;25:431–6.

Chapter 8

Pages 32–35, "If the Normal Distribution is So Normal, Then How Come My Data Never Are?" *Medscape*; WebMD, 111 8th Avenue, New York, NY 10011.

Chapter 11

Pages 46–49, "Statistical Ties and Why You Shouldn't Wear One." *Medscape*; WebMD, 111 8th Avenue, New York, NY 10011.

Chapter 13

Pages 54–56, "To *p* or Not to *p*." *Medscape*; WebMD, 111 8th Avenue, New York, NY 10011.

Chapter 14

Pages 58–59, Petrosino, Anthony et al. (2005). "Scared straight and other juvenile awareness programs" in B. Welsh and D. Farrington (Eds.), *Preventing crime: what works for children, offenders, victims and places* (pp. 87–101). The Netherlands: Springer. With kind permission from Springer Science and Business Media.

Chapter 15

Page 62, "Low-Fat Diets Flub a Test", Editorial, *New York Times*, 9 February 2006.

Chapter 17

Page 75, Reference to "sample size samba." Reprinted from Schultz, K. and Grimes, D., "Sample Size Calculations in Randomised Trials: Mandatory and Mystical." *The Lancet*, Vol. 365, Pages 1348–1353, © 2005, with permission from Elsevier.

Chapter 19

Pages 86 and 87, Citation and discussion of study about fast food and exercise habits; "Eating More, Enjoying Less." *Pew Resarch*. published April 19, 2006. ©2009 Pew Research Center, Social and Demographics Project. www.socialtrends.org.

Chapter 22

Pages 99–100, "Three Statistical Errors That Are Totally Trivial but Which Matter a Great Deal." *Medscape*; WebMD, 111 8th Avenue, New York, NY 10011.

Chapter 24

Page 107, Data, tables, and discussion regarding interfaith marriage survey. *Northern Ireland Life and Times*, ARK, School of Sociology, Social Policy and Social Work, Queen's University Belfast, University Road, Belfast, Northern Ireland. BT7 1NN.

Chapter 26

Pages 116–118, "Regression to the Mike: a statistical explanation of why an eligible friend of mine is still single (and some implications for medical research)." *Medscape*; WebMD, 111 8th Avenue, New York, NY 10011.

Chapter 28

Pages 127–128, Citation and discussion of study regarding use of aspirin after a heart attack analyzed by astrological sign. Schultz, Kenneth F. and Grimes, David A., (2005, April). "Sample size calculations in randomized trials: mandatory and mystical." *The Lancet, 365*(9467): 1348–1353.

Chapter 30

Pages 133–137, "How to Win the Marathon: A Common Statistical Error Can Help." *Medscape*; WebMD, 111 8th Avenue, New York, NY 10011.

Page 135, Citation and discussion of study re: survival rates in lung cancer patients diagnosed by CT scan. Bach, Peter B. et al, (2007, March). "Computed Tomography Screening and Lung Cancer Outcomes." *Journal of the American Medical Association, 297*(9): 953–961.

Chapter 32

Pages 142–146, "Statistics Is Unscientific! (Well, as Clinicians See it, Anyway)." *Medscape*; WebMD, 111 8th Avenue, New York, NY 10011.

Chapter 33

Pages 147–150, "Numbers That Mean Something: Linking Math and Science." *Medscape*; WebMD, 111 8th Avenue, New York, NY 10011.

Page 148, Portrait of the English statistician and geneticist, Sir Ronald Aylmer Fisher, ©Photo Researchers, Inc.

Chapter 34

Pages 151–153, "Statistics Is About People, Even If You Can't See The Tears." *Medscape*; WebMD, 111 8th Avenue, New York, NY 10011.

Page 152, Map: public domain.

Discussion Section

Chapter 6

Page 162, Bar graph: length of paid maternity leave in different countries. Gornick, Janet C. and Marcia K. Meyers. Figure 5.2, "Paid Leave Available to Mothers, Approximately 2000." In *Families That Work: Policies for Reconciling Parenthood and Employment.* © 2003 Russell Sage Foundation, 112 East 64th Street, New York, NY 10021. *Reprinted with permission.*

Chapter 9

Page 166, "Great Male Survey: 'our results reveal who the modern man is' and '56% of men see the drink they order as a reflection of their masculinity or character.'" *AskMen.com.*

Chapter 24

Page 190, Data used in graph showing crime data relative to poverty rate. StatCrunch/DOJ.
Page 191, Data used in graph showing crime data relative to metropolitan vs. rural areas. *Ibid.*

Chapter 27

Page 198, Shakespeare's *Othello*.

Chapter 32

Page 205, 19 lines of code from software package, printout for *t* test. Stata Statistical Software. College Station, TX: StataCorp LP.
Page 205, line of programming code. "The difference between groups is 0.6. The 95% confidence level is 0.1 to 1.2." Stata Statistical Software. College Station, TX: StataCorp LP.

Index